Science, Society and Sustainability

Routledge Research in Education

Science, Society and Sustainability

Education and Empowerment for an Uncertain World

Edited by Donald Gray, Laura Colucci-Gray and Elena Camino

Routledge
Taylor & Francis Group

New York   London

First published 2009
by Routledge
711 Third Avenue, New York, NY 10017

Simultaneously published in the UK
by Routledge
2 Park Square, Milton Park, Abingdon, Oxon OX14 4RN

Routledge is an imprint of the Taylor & Francis Group, an informa business

First issued in paperback 2011

© 2009 Taylor & Francis

Typeset in Sabon by IBT Global.

Library of Congress Cataloging in Publication Data

Science, society, and sustainability : education and empowerment for an uncertain world / edited by Donald Gray, Laura Colucci-Gray, and Elena Camino.
    p. cm. — (Routledge research in education ; 27)
    Includes bibliographical references and index.
 1. Science—Social aspects.   2. Science—Study and teaching—Social aspects.
3. Research—Social aspects. I. Gray, Donald, 1955– II. Colucci-Gray, Laura. III. Camino, Elena.
    Q175.5.S3724 2009
    303.48'3—dc22
    2009000559

ISBN13: 978-0-415-89795-2 (pbk)
ISBN13: 978-0-415-99595-5 (hbk)
ISBN13: 978-0-203-87512-4 (ebk)

To children and grandchildren everywhere, and to natural systems that sustain our lives.

Table of Contents

PART I
The Changing Aspects and Roles of Science:
Implications for a Sustainable and Democratic Society

PART II
Science and Sustainability:
Implications for the Learning and Teaching Process

Boxes

Figures

Plates

Tables

Preface

This book comes at a time when the education community—and particularly science education, in Europe and beyond—is questioning its aims and practices to meet the needs of young citizens. Much debate has concerned the relevance and appeal of science education for young people[1]; more generally, UNESCO and the United Nations have called for an education that can respond to the needs of a world that is increasingly more interconnected, complex and characterized by conflict.

This is the context in which this book originated: as a group of academics and practitioners we have tried to enter such an arena of multiplicity and idiosyncrasy, with the purpose of learning something and changing something of ourselves and our practice along the way.

Over the course of the years we have come together as a group of people from different backgrounds, nationalities and expertise that have been grappling with the problems and issues of science and sustainability. All of us have engaged with science in our careers and many of us hold a science degree. However, at different points in our professions we found ourselves in the position of questioning our knowledge, our learning and our understanding of the complex problems of our time and started to become involved with possibilities for change. This book is intended to provide an overview of this long-standing inquiry, about science, sustainability and educational practices, drawing on the variety and multiplicity of our experiences as researchers and educators in lower and higher education.

Issues of sustainability are interdisciplinary in nature, involving a multiplicity of fields of inquiry. In this book, we put forth the notion of dialog (between disciplines, forms of knowledge and different points of view) as a methodological approach to deal with the complexity of the questions we face. Central to our discussions are also the notions of complex systems and nonviolent conflict transformation, which are powerful frames for changing from an environment of hierarchical, asymmetrical relationships to a context of equality, participation, engagement and mutual recognition of interdependencies. Yet, with dialog also comes diversity, otherness, puzzlement and discomfort. As will be better defined in the section that follows, the authors of this book make for an extremely variegated set of biographies and work geographies. A framework of sustainability has brought many of us together to form a group of people interested in collective, interdisciplinary inquiries[2]. In this collaborative work we have found ourselves stuck many times in trying to enter other people's ways of perceiving topics and problems, which were at times buried under the long-winded sentences of Italian writers or the visually dense expressions of the English contributors. So, more than once we had to take stock and step back: to rethink what we wanted to say and take a different perspective. In this way we have endeavored to touch upon very strongly consolidated ideas about science and about education while trying to preserve the pleasure of discovery and surprise.

We propose to engage with the readers in a similar fashion, using the book as a tool for reflection on the many issues surrounding the controversial and sometimes paradoxical relationship between science and society. The specialist reader can find the opportunity to engage with case-study analysis and meta-reflections.

The critical educator and critical reader can draw upon some of the research-laden approaches to engage with a vision and a practice of learning and knowing that is in line with the complex and evolving scenario of sustainability.

In this regard, a final thought needs to be given to the use of vignettes. At different points throughout the book we have inserted vignettes produced by Massimo Battaglia, one of our colleagues in the Inter-university Institute for Research on Sustainability. Using his artistic talent, he tried to represent some of the concepts that have characterized the discussions of the group. The result is a series of drawings (of which the book contains a selected number) which are offered to the readers as a tool for engaging with personal interpretation, using a multiplicity of channels, from the cognitive to the emotional, as part of this exploration of the landscape of sustainability.

REFERENCES

Osborne, J. and J. Dillon. 2008. *Science education in Europe: Critical reflections: A Report to the Nuffield Foundation.* London: Nuffield Foundation.
OECD. 2006. *Evolution of sudent interest in science and technology studies.* Policy Report. Paris: OECD.
EURYDICE. 2006. *Science teaching in schools in Europe: Policies and research.* Brussels: European Commission.

NOTES

1. See, for example, Osborne and Dillon 2008; TLRP 2006; OECD 2006; EURYDICE 2006.
2. Many of us are members of the Inter-university Institute for Research on Sustainability based at Turin University (www.iris.unito.it).

Acknowledgments

We would like to thank the Assessorato Ambiente of the Piedmont Region, Italy, which over the years has supported much of the research undertaken by the Interdisciplinary Research Institute on Sustainability (www.iris.unito. it). Also, the work of the science education group at the University of Turin would not have been possible without the MURST[1] funding held by Elena Camino and Anna Perazzone. We would also like to thank the College of Arts and Social Sciences at the University of Aberdeen, which provided the pump-priming funds that allowed the editors to meet and take this project forward, as well as enabling dialog and other collaborations to take place.

A special mention should be made of Laura Colucci-Gray who, babe in arms, toiled relentlessly to translate the four chapters in this book that were originally written in Italian, as well as contributing to all the other editorial requirements.

We would also like to thank Massimo Battaglia for generously allowing some of his vignettes to be reproduced in this book.

NOTES

1. Ministero per l'Università e la Ricerca Scientifica e Technologica.

Part I

The Changing Aspects and Roles of Science

Implications for a Sustainable and Democratic Society

Introduction to Part I
Issues and Scenarios

*Donald Gray, Laura Colucci-Gray
and Elena Camino*

A NEED FOR A MULTI-LEVEL CHANGE

The world as we know it is changing. Daily we read of global financial meltdowns, of climate change, and social and ecological disasters. While there have been dramatic improvements in health and welfare and in material standards of living (at least for a part of humanity), in communication and entertainment technology, somehow the world is beginning to look less secure, and the speed of this change is accelerating.

This book was born out of a concern about these developments and a conviction that these changes are being brought about partly because of an inherent problem—particularly in Western and westernized societies—with dominant ways of thinking about the world, how it functions and the relationships between human societies and natural systems that support them. Scientific research and science education can play a fundamental role in critically reflecting on such dominant ways: in this sense, these two areas of human inquiry constitute the two strong focuses of reflection in this book.

This is a book written largely by people with backgrounds in science, some working in science departments in a university and others working in science and teacher education. All the authors have engaged with a reflection on science and science–society interactions as part of the question of how human societies can move towards more sustainable relationships with the environment and other living beings.

There are two key strands that run through this book. One is the theoretical foundation of *sustainability science*[1], which reconsiders both the epistemological assumptions underpinning scientific research and the relationships between science and society to provide a coherent framework for accounting for the complexity and change characterizing our time. The other is the theoretical framework of sustainability education, which is presented as a series of examples of applications of the ideas of sustainability science within educational contexts. The book is organized in two parts that focus on these two strands, although the ideas in the chapters are not a simple linear flow and there are many links across the different chapters at different parts of the book. We have not tried to eliminate possible overlaps, but merely tried to minimize unnecessary repetition, and the ordering of the chapters is the recommended order of reading. For those who are more familiar with the ideas presented here it is also possible to pick up and start from any point, the common theoretical ideas being sufficiently highlighted in each chapter to enable the connections to be made.

All the authors agree upon a number of aspects being of the greatest importance for reflection on science and education: the growing awareness of complexity and uncertainty in the realms of science, society and environment interactions; the greatly accelerated and unsustainable pace of change as a result of the power of science and technology to transform and manipulate the resources of a planet that is bounded and finite; the nature of current knowledge production and use and its social implications (including the rise of conflict); and finally, the need to expand participation and decision-making processes with respect to complex socioenvironmental issues. Such aspects represent the conceptual framework of the entire book. In the face of these concerns, there is a need to address the type of thinking that is offered and spread at all levels of education: thinking that is predominantly shaped by a positivistic epistemology, a mechanistic linear model of cause and effect, and a methodological approach based on reductionism. In order to embrace complexity and uncertainty a more holistic, systemic approach to science and science education is required. This approach implies an appreciation and understanding of the limits of disciplinary knowledge as a particular and bounded perspective; an understanding of the importance of the relationships amongst parts of a system, which are not captured by analytical and quantitative methodologies; and a more inclusive approach to different types

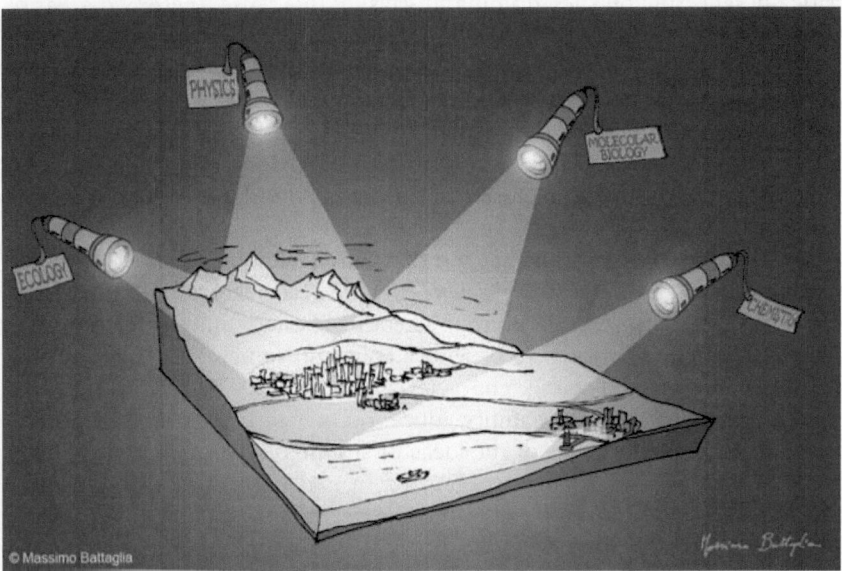

Figure I.1   Perspectives and uncertainty.

of knowledge, of which scientific knowledge is only one. A more fluid and inclusive way of thinking asks for putting together reflection and action: different patterns of thinking, mindsets and mental schemes can be reframed within practices that do not separate and divide but which seek to relate and put into dialog. This is change at multiple levels, from the wider social context to the more personal interactions of educational encounters, as will be detailed in the section which follows.

CONTENTS AND ISSUES OF THE CHAPTERS

Part I: The Roles of Science and Implications for a Sustainable and Democratic Society

In the first chapter, Alice Benessia begins with an observation of the changes that have characterized modern scientific research and introduces us to the relationships between technoscientific knowledge and socioenvironmental problems. From an examination of complexity, uncertainty and irreversibility of many science–society–environment interactions she introduces the idea of *indeterminacy* arising from competition between different disciplines and the social, political and ethical interactions. Abandoning the notion of "science speaking truth to power," Benessia leads us on to new thinking in "post-normal" science, developed by Funtowicz and Ravetz, as a framework for thinking about science, knowledge production and decision making in a scenario of increasing uncertainty where decision stakes are high.

The scenario of post-normal science is picked up in Chapter 2, where Ângela Guimarães Pereira elaborates on models of democratic decision making in the science–policy interface. She starts from the concept of the Public Understanding of Science, which was conceived on the idea that public support, or lack thereof, for science was simply related to the level of the population's scientific literacy and that a public better educated to understand science would be more supportive of science. Guimarães Pereira provides a coherent account of an evolutionary perspective of the governance of science based on the "extended participation" model inherent in post-normal science, offering an adequate context for new relations between science and the public. Continuing the exploration, Guimarães Pereira goes on to elaborate on "civic epistemologies" and knowledge co-production, based on the concept of "extended peer review." She concludes the chapter by suggesting that "we have to learn to think in a new way" and that rather than science (unconsciously) adopting an arrogant approach to other knowledges, there is a need for humility in dealing with "non-scientific" views and in acting on natural systems.

In Chapter 3, Luca Giunti and Elena Camino build on some of the issues explored in the first two chapters by taking a close look at the particular case of the proposal for the construction of a high-speed rail link through a mountain valley. The authors paint a picture of a situation that consists of a "complex weave of worldviews, points of view, interests, needs which [have] developed over the years around a complex and controversial socio-environmental issue" (Giunti and Camino, Chapter 3, 68). The case examines conflicts of views between political and economic groups at a national level on the one hand, and the local population on the other; it exemplifies dynamics and problems that are common to many issues in science and society, raising important questions about how to take forward and address issues of genuine citizens' participation in matters of public life. In this regard, the case of the Susa Valley explores the implications of a nonviolent approach in seeking participation, dialog and a creative transformation of the conflict.

Finally, Chapter 4 looks at language: this is the medium through which communication occurs, but it is also a tool for producing new ideas. Starting with a semiotic analysis, Dodman and Camino lead us through the structure of language and its different forms and ways of describing reality. This highlights the dynamic nature of language as an open, continuously evolving system in itself. Through a careful analysis of nature and use of language, Dodman and Camino offer some cues about the implications of practicing different kinds of language, and about the role of language—in particular, scientific language—in shaping power relations within a society. They argue that promoting awareness of the role of language in building new knowledge and multiple world views should be a primary objective both for science and for science education.

In this chapter, the author of Chapter 1, Alice Benessia, offers an integrative reflection about visual language in science: different ideas of science are "illustrated" by a different iconic language that can be used for persuasion about "truth" as well as for developing critical thinking.

So, the first part of the book concludes with the profile of a framework for looking at science and society relationships. Participation and dialog are central to decision making and knowledge building and they are the methodological choices that are put forward to build awareness of the complexity of the reality in which we are immersed. Similarly, such understanding can give pointers for an education that seeks to prepare people to act within a changing epistemological condition.

Part II: The Implications for Science Education

From an exploration and analysis of science and socioenvironmental issues in the first part of the book, the second part considers the role of education in taking forward a transformative agenda for science, society and sustainability.

The second part begins with Chapter 5, in which Stephen Sterling points to the epistemological assumptions of current education, which is defined as "of a previous age." This is an age that was dominated by a mechanistic outlook on the world, according to the idea of an ordered reality that obeys universal laws. Yet it is a view that is at odds with a systemic perspective of complexity and uncertainty, derived from an exploration of living and evolving processes. Sterling argues for a transformative education that he calls *sustainable education*, which is different from, but contains, *sustainability education*, and requires a change in the culture of education as a whole. This, he argues, requires a fundamental change in the epistemology of our culture and thus in educational thinking and practice. Sterling, using the metaphor of a holistic medical practitioner, suggests focusing on the subject rather than on the symptoms of illness, investigating the kind of society, or mindset, that has produced unsustainable situations, and questioning how and why current thinking and educational practices have created fragmentation and separateness between us and the world around us. In order to remedy this there is a need for:

> a shift of emphasis from relationships largely based on separation, control, manipulation, individualism and excessive competition towards those based on participation, appreciation, self-organisation, equity, justice, sufficiency and community. (108)

From this premise, Sterling goes on to construct a vision of what a sustainable education system would look like, which entails a shift away from thinking about how do we educate for sustainability to a much deeper focus on education itself; in other words, "its paradigms, policies, purposes, and practices and its *adequacy* for the age we find ourselves in." (111)

From a consideration of what thinking means for education as a whole, in terms of systems, complexity and uncertainty, in Chapter 6, Camino, Barbiero and Marchetti provide a closer look at some examples of how such thinking might be applied in two specific contexts in higher education. One examines a purpose built course provided for an interdisciplinary group of doctoral students; the others refer to approaches used in initial training courses for future secondary school science teachers. The chapter begins by recognizing that courses for future science teachers and future researchers are characterized—at least in Italy—by the accumulation of specialized disciplinary knowledge and by traditional transmissive modes of teaching. As it is clearly acknowledged by science education research, such a focus and such an approach results in fragmented knowledge, poor critical thinking and a fracture between "experts" on one side, with privileged status with respect to knowledge claims, and lay people on the other, often disempowered in decision-making processes related to socioenvironmental issues. Such developments are incompatible with the ideas of complexity and uncertainty in a post-normal science framework, where there is a need for consideration and acknowledgment of multiple perspectives and others' views, and a requirement for an understanding of the systemic relationships inherent in socioenvironmental contexts. This reflects Sterling's standpoint in Chapter 5 that education systems today are still founded on outdated epistemological foundations and there is a need for a more holistic approach to education. The courses developed and reflected on in this chapter by Camino, Barbiero and Marchetti exemplify an epistemological and methodological approach, grounded in the development of *conceptual tools,* which is aimed at a transformation from a reductionist and objectifying view towards a holistic and integrated view. A goal of these courses is to provide young people with the skills to participate in an extended democracy that is more inclusive of wider knowledge claims and value perspectives. The strategies used in the teaching/learning context with these groups tend to encourage and promote a transformation from previous hierarchical structures and unbalanced power relations to more balanced relational approaches where people have equivalent value and each one can legitimately express her/his point of view. Relationships are gradually transformed from those of confrontation and conflict to mutual understanding and dialog. Such a different attitude is fundamental in engaging in nonviolent conflict transformation where views and opinions may be divided, as they often are in socioenvironmental contexts.

Moving from secondary education in Chapter 6 to primary education in Chapter 7, Angelotti, Perazzone, Tonon and Bertolino provide a description of how the principles discussed in the previous chapters have been applied to training courses for future primary school teachers, and present an analysis of the role that primary education has in relation to environmental education and education for sustainability. The authors of this chapter provide a comprehensive overview of the nature of primary teacher education in Italy, which is not dissimilar to that in many other Western countries, and consider the constraints and opportunities that exist in their context for taking forward some courses aimed at developing competencies in education for sustainability in future primary teachers. They consider environmental education more as a way of promoting new ways of thinking, educating and relating to others rather than as a form of knowledge transmission. They suggest that the "revolutionary" ideas of Sterling (Chapter 5) in this context can only be achieved through transdisciplinary visions that "originate from a convergence of objectives and a concerted use of interpretive schemes that are suitable for overcoming the cognitive obstacles preventing sustainable living." (158) According to the authors, there is a need to develop approaches that can reorient students to a perception of themselves as part of that natural world and that include knowledge not as a "scholastic object," but rather as a means of inclusion and orientation within the natural systems. Taking a complementary approach and drawing on the perspective of biophilia, Giuseppe Barbiero describes activities, such as silence and meditation, used to reconnect children's affiliation with the natural world.

Each of the chapters in Part 2 of the book has been concerned with the development of approaches to a form of education that is transformative and aimed at shaping thinking and practices for moving towards a more sustainable future. Laura Colucci-Gray, in Chapter 8, continues along the same track by considering a case study of pedagogical practice. Building on ideas contained within Chapters 1 and 2 with respect to post-normal science and participatory processes, and incorporating pedagogical approaches mentioned in Chapters 6 and 7, Colucci-Gray focuses on a case study of the issue of prawn farming in India used as the context for a role-play undertaken by secondary school science students. The chapter starts by exploring the choices of content and methods that might be made to promote a "sustainable" education, settling on the proposition that role-play is an appropriate pedagogical and conceptual tool for creating contexts in which students can develop competencies for dealing with complex global controversies and conflicts. Drawing on the works of other scholars in the field of science studies, Colucci-Gray elaborates on the source of many conflicts as originating in the clash of perspectives between so-called indigenous knowledge and the interventions of so-called expert knowledge. Taking this argument further results in a justification for role-play as a way of developing an empathetic understanding of multiple points of view. However, rather than settling on role-play as it is often conceived in an educational setting as a simulation of a decision-making process resulting in a judgement, Colucci-Gray takes the process a step further by engaging previously antagonistic groups in a process of nonviolent conflict transformation to arrive at a consensus for action towards a sustainable future. Thus, Colucci-Gray takes us on a journey of exploration in role-play that can, in her words, "be seen from different perspectives: just as a teaching tool or as a means for participation, engagement and development of new competences." (207) The perspective taken here is that, used in this way, role-play is one approach to providing a solid pedagogical approach to the development of citizens with appropriate competences for engaging in science and policy debate in a post-normal scenario.

By its very nature, the process of social and cultural change supported by a vision of sustainability requires the involvement of many actors: from politicians to citizens, teachers and children in school. Our work has focused largely on our experiences with beginning teachers in higher education with whom we endeavored to share what has been an evolving reflection on aims and methods of a sustainable education in a cultural and political context that has not always been favorable. Hence, in approaching the reading of this book, we hope that the activities and reflections reported here can become stimuli for further inquiry, a contribution toward the search for a more harmonious and sustainable future.

NOTES

1. As defined by Gallopin et al. (2001) throughout the book.

BIBLIOGRAPHY

Gallopin, G.C., S. Funtowicz, M. O'Connor, and J. Ravetz. 2001. Science for the 21st century. From social contract to the scientific core. *International Social Science Journal* 53 (168) : 219–31.

1 From Certainty to Complexity
Science and Technology in a Democratic Society

Alice Benessia

EDITORS' INTRODUCTORY NOTES

This chapter begins the reflection—central to Part 1 of the book—on the changing relationships between science, technology and society, which characterize current issues of sustainability at both the local and global levels. More specifically, the aim of this chapter is to introduce and problematize the concept of technoscientific knowledge in socioenvironmental problems and its role in normative frameworks, providing a general overview of the most recent developments in the area of science and public policy.

From a deterministic perspective to scientific knowledge, uncertainty is considered marginal and temporary: in the long run, science is thought to provide certain and objective sets of facts, independently validated by a community of peers, and "to speak truth to power," contributing through that process to the normative sphere.

However, recent years have provided the context for a profound epistemological shift: complexity, irreversibility and indeterminacy undermine the idea that science can provide single, objective and exhaustive answers, so that the right course of action can be settled. This chapter introduces the framework of post-normal science, in which the ideal of a rigorous scientific explanation is replaced with the ideal of open public dialog, introducing the epistemic and methodological necessity to include public participation in decision-making processes on socioenvironmental issues.

The chapter ends by introducing the thinking, which will be deepened throughout the book, about the profound implications that such transformations of the idea and role of science can have for both civic life and education: the idea that dominant narratives have to be complemented with a plurality of approaches, methodologies, languages and ways of looking at reality that are democratically elaborated and shared.

THE CHANGING NATURE OF SCIENCE–SOCIETY INTERACTIONS

In the last century, science and technology have gained an unprecedented influence on society and the environment. What is remarkable is their severely intense capacity to transform, manipulate and transfer energy and matter over local and global scales, and shorter and shorter time scales. In more general terms, we are now facing a period of exponential growth in the power of human socioeconomic systems to transform natural systems, and this growth has been boosted by important changes in the nature of scientific enterprise. The disciplinary boundaries and the once fundamental distinction between "pure" science and "applied" technology are merging into hybrid, upstream *technosciences*. Defined by the acronym GRAINN by Ravetz (2006), these include fields such as genomics, robotics, artificial intelligence, neuroscience and nanotechnology, which, in contrast to more traditional forms of knowledge generation in science, are characterized by research and development strategies targeting the creation of wealth and products within a globally competitive economy.

Hence, technosciences are sophisticated tools that are deeply embedded in the economic life of a nation, and this new face of scientific research has a number of implications. Progressive concentration of private financial, economic and political power is associated with producing, managing and implementing scientific and technological research. In addition, public knowledge is transformed into private and corporate know-how (i.e., through the mechanism of patenting). The modern isolation of scientific research from the political and normative arena has been deeply undermined, as technosciences influence the sociopolitical decisions of a nation. It is this relationship between science and public policy that is made problematic in this chapter.

In the last sixty years, traditional disciplinary, curiosity-oriented science, such as physics, chemistry and, more recently, biology, once regarded as solely and inherently beneficial for human kind and the environment, exempted from the fallacy of human institutions and associated with the myth of inexorable progress, have come to be overtly controversial. They are controversial because of their military applications, such as the atomic bomb, the Nazi gas chambers and the more recent Agent Orange in Vietnam. Because of the unpredictable side effects of their implementations, such as those associated with the green revolution, as exposed in the famous Rachel Carson best-seller, *Silent Spring*, or the most recent BSE or "mad cow" syndrome in England. They are also controversial for the numerous and shocking "normal accidents" (as perceptively defined back in 1984 by the sociologist Charles Perrow) such as the Chernobyl and the Bhopal disasters (Perrow 1984). All such examples point to a scientific enterprise that is context-based, shaping and, at the same time, shaped by the culture and values of a society. Quite clearly, all these cases could not be predicted nor confronted in an isolated manner: sets of values and specific choices were deeply embedded in the outcomes. Besides, the quantitative and specialized knowledge of disciplinary science could not be considered as neutral. In this way, technoscientific research continuously redefines citizens' lifestyles and value systems, but it also involves risks and uncertainties that cannot remain unaddressed. In addition, the emergence of global socioenvironmental issues, such as the ozone hole, biodiversity decline and climate change, have raised the awareness of the impact of our technoscientifically driven ways of life on the planet, and of the need for new global values and normative actions (Singer 2002).

THE CONTRIBUTION OF THIS CHAPTER

This chapter explores a field of reflection on the need for new social and political structures and methodologies through which public space could be opened in the decision-making processes involving science and technology. However, it is argued that in order for the extended participation to be effective and intelligent, more attention needs to be given to raising public awareness and empowering individual and collective imaginations. In other words, we have to enhance our capacity to creatively reflect on the interface between the factual approach, based on "what we know" and "what we can do," and the normative framework, founded on "what we want" and "what we need." As we will see in this chapter, and more specifically in Chapters 3 and 4, a number of specific and powerful master narratives are embedded in our culture and our society in relation to science and technology. These narratives are the expression of wider imaginations about our world, what we value and what we need to achieve our aims. A new social contract between science, policy and citizens, then, has to be based on a conscious evaluation of these visions and narratives on the one hand, and on the collective reflection and conception of new ones, oriented towards the creation of more sustainable futures, on the other hand.

In what follows, I will first give a description of current master narratives and how these have become critical in current times. I will then describe the features of a changing normative framework in which citizens can be empowered to implement choices and decisions that are respectful of the complexity of socioecological systems and can support a view of sustainability.

FROM EXPERIMENT TO EXPERIENCE: THE MODERN MODEL AND ITS DEVELOPMENTS

The question we urgently face is how to live democratically and at peace with the knowledge that our societies are inevitably at risk. (Jasanoff 2003, 224)

One of the main features of contemporary high-power technoscience is that research and implementation are undertaken outside of the secured, controlled and simplified setting of the laboratories, and directly experimented—or we should say *experienced*—on the socioenvironmental systems of the planet. This major transition implies at least three orders of consequences, which in turn entail that we radically rethink the relationship between our capacity to understand and manipulate social and natural systems and our ways of acting upon this knowledge and power.

The first consequence is that the *complexity* involved in the interaction between the socioenvironmental systems in which the implementation occurs and the technology itself is radically different from the mere simplifiable complication of laboratory settings. Therefore, as we will see, uncertainty and ignorance are intrinsic and unavoidable.

The second consequence is that direct experimentation is *not reversible*: there is no way to run a second test—that is, to have a second chance—when natural systems are involved. All together, these two factors imply that negative, unforeseen, unpredictable and unrecoverable outcomes can occur. In other words, technoscience ensures benefits, and at the same time it generates risks that cannot necessarily be addressed by technoscience alone. This scenario is what the sociologist Ulrich Beck defines as the *risk society* (Beck 1992): risks are endemic in technoscientific-oriented contemporary societies. In Sheila Jasanoff's words, they are "part of the modern human condition, woven into the very fabric of progress" (Jasanoff 2003).

The third order of consequences lies in the fact that this kind of open-field implementation doesn't depend on a set of disciplinary methodological choices, embedded in a specific experimental setting. Hence, the kind of knowledge that is needed to describe and deal with the socioenvironmental impact of the development and the introduction of a new high-power technology is not defined a priori; it is the result of a series of choices, of a negotiation and more often of a competition between different disciplines. This factor introduces a level of *indeterminacy* (Smith and Wynne 1989) in the very definition of the issues to face, in terms of disciplinary and methodological options. As Daniel Sarewitz puts it: "The necessity of looking at nature through a variety of disciplinary lenses brings with it a variety of normative lenses, as well" (Sarewitz 2004, 385).

Technoscience and the Normative Sphere

Innovation Science

Given the major transition that we have so far outlined, the relation between technoscience and the normative area of ethics, politics and the law becomes more and more relevant and intricate at the same time. On the one side, technoscience is the *object* of politics: someone has to decide where, how, when and, regrettably rarely, *why*, to implement it—that is, to regulate its development and applications. On the other side, scientists are called to *inform* decision makers, to provide a general overview, the state-of-affairs about the interaction between the socioenvironmental systems and the technology under examination. Most of the time, these two functions are strictly correlated: the same scientific knowledge that emerges from the development of a technology is employed to regulate the technology itself. The case of genetically modified organisms is emblematic in this respect.

Corresponding to these two functions are two different ways of conducting scientific research: the so-called *innovation science* (Wynne, quoted in Jasanoff 2005), in charge of producing new technological products, and *regulatory science* (Jasanoff 1990), responsible for managing and restraining their applications. Innovation science is based on controlling and programming specific, profitable and advantageous properties for who produces the technology and possibly for who makes use of it. The aim of innovation scientists is reflected in their methodologies, essentially based on linear cause–effect relations that occur between a limited, or at least limitable, number of variables. The theoretical frameworks that best suit this kind of approach are reductionism and mechanism: the behavior of a system can be predicted and controlled by reducing it to a fundamental set of subsystems, from which the dynamics of the whole can be analytically deduced.

Regulatory Science

Regulatory science is essentially founded on the same approach as innovation science, and sometimes it is also run by the same people. In both cases, the provisional character of scientific knowledge—ignorance, uncertainty, indeterminacy, disagreement and surprise—is eliminated by closure mechanisms: in the first case, by applying the approach of deterministic control through quantitative predictive technologies; in the second, by legitimizing a given body of knowledge through the institution of "expert committees," appointed by decision makers. In order to be translated in normative form and to be implemented as market products, technoscientific knowledge indeed has to be definitive. An illuminating example of this procedure of closure—a case of what Jasanoff refers to as "co-production" of epistemic and normative elements (Jasanoff 1995)—is the debate about the definition and the function of genes in relation to the issue of patenting simple and then complex genetically modified organisms. This particular example shows that in both the *innovation* and the *regulatory* sectors, the reductionist, mechanistic and deterministic approach that associates one gene to one protein and one function had to prevail over the equally scientifically proven recognition that groups of genes can oversee the expression of different proteins with multiple, unpredictable effects. Hence, the process of genetic engineering of biological organisms cannot be subjected to simple and univocal normative actions of general value.

These two kinds of technoscientific research are based on the modern ideal of scientific knowledge, characterized by certainty, objectivity and the capacity to exhaustively and univocally describe the reality under examination. In the innovation science approach, uncertainty is considered as marginal, temporary and confinable; the possible outcomes are described in quantitative terms and out-of-norm events are dealt with by a posteriori correction procedures, with the tendency to extend the *trial-and-error* mechanism outside of the controlled and reversible setting of the laboratory. In regulatory science, the debate about uncertainty is closed by a normative action, maintaining the discussion in the certain, objective and exhaustive parameters of the modern ideal.

In this model of knowledge production, the "good" of politics derives automatically from the "truth" of science: in other words, science is in charge of providing the certain and objective set of facts from which the right course of action rationally derives. In Wildavky's words: "science speaks truth to power" (Wildavsky 1979). In this ideal, defined by Merton as the "republic of science" (Merton 1968), scientific knowledge is provided with a privileged epistemological status, as it is produced by a community of peers, self-governed without any form of coercion or authority other than knowledge itself. The normative sphere is then conceived of as a mechanical application of a kind of knowledge that is validated elsewhere: facts and values are independent and separate.

It is clear from this overall picture that there is a deep gap between the methods and the assumptions of both innovation science and regulatory science, the first attached to the modern model from which it stems; the second elaborated in authoritative defense of the model itself, and the actual scenario in which technoscience, society and the environment are embedded. More specifically, as we have seen, complexity, irreversibility and indeterminacy radically undermine the idea that science can provide a single, certain, objective and exhaustive perspective from which a straightforward decision can be taken at a political level.

As we will see also in the following chapter, in the last twenty years, new models of interaction between science and the normative sphere of policy and the law have emerged in order to cope with these deep changes.

Dealing with Uncertainty

The first acknowledgment of the actual lack of knowledge embedded in the decision-making process regarding socioenvironmental issues, and of the necessity to actively deal with the irreversibility of technoscientific high-power experimentation, was provided in 1992, in Principle 15 of the Rio Declaration on Environment and Development:

> In order to protect the environment the precautionary approach shall be widely applied by States according to their capabilities. Where there are threats of serious and irreversible damage, lack of full scientific certainty shall not be used as a reason for postponing cost-effective measures to prevent environment degradation.

The precautionary approach introduces the idea that science can be *temporarily* unable to produce a certain and exhaustive body of knowledge suitable for a rational decision. In that case, a political principle has to be instituted in order to minimize the chance of harming people and the environment over the chance of an economic loss due to a technological restriction. In other words, whenever full knowledge is lacking, a political choice is made in order to minimize the so-called type-II errors; that is, the errors of accepting harmful developments (false positives) over the typical progress-based optimistic ideal of minimizing type-I errors, corresponding to the errors of rejecting harmless developments (false negatives). The fundamental consequence of this normative step is an epistemic shift from a two values theoretical science, based on the evaluation of the truth/falsity of a hypothesis, to a three-value science "applied to risks", which includes uncertainty as a possible outcome of a scientific assessment. This in turn implies a deep change in the modern relationship between scientific truth and the correspondingly right political decision by introducing the notion of risk *acceptability* through a normative action (Tallacchini 2005; Shrader-Frechette 1996). Indeed, in the 2000 Communication from the European Commission, the precautionary principle is qualified as a principle of political responsibility, namely a principle that considers certain risks as "inconsistent with the high level of protection chosen for the Community" (Communication from the European Commission 2000).

The precautionary model represents a substantial improvement over the inherent positivism of the modern ideal, in that uncertainty and the need for an eminently political stand are explicitly taken into account. Nonetheless, the privileged nature of scientific knowledge is not challenged at its roots: lack of full knowledge is still regarded as technical uncertainty, a provisional condition ascribable to temporary methodological difficulties in collecting and managing data. Moreover, in the same Communication from the European Commission, the precautionary principle is associated with risk management: it can be invoked only when a scientific assessment provides evidence of risk and when the precautionary measures are in line with a proportionality principle between cost and benefits (Funtowicz 2007). In other words, in this model, lack of knowledge is *reduced* to a temporary and statistically manageable condition.

Another substantial limitation of the precautionary principle lies in its philosophical foundation, based on Hans Jonas *heuristic of fear* (Jonas 1985). According to his perspective, when full knowledge is lacking, it is more responsible to believe in the prophecy of doom than in encouraging predictions. In other words, it is wiser to prioritize fear over hope. This implies that the unknown is not provided with an epistemic, but rather a *psychological* status, and that the modern positivistic dichotomy that associates full and certain knowledge with rationality on one side and the unknown with the erratic world of emotions on the other, is reconfirmed. Moreover, this reference to the realm of the irrational leads part of the scientific community to reject precaution as a form of obscurantism.

Given the increasingly evident lack of agreement in scientific assessments, not only certainty but also the two other pillars of the modern ideal, objectivity and exhaustiveness, have been deeply challenged in recent times. Correspondingly, as we will specify in the next chapter, two different models of interaction between science and policy have been elaborated in Europe: the *framing* model, which specifically acknowledges indeterminacy, and the *demarcation* model, which deals with the issue of conflict of interest (Funtowicz 2006). More specifically, in the *framing* model all stakeholders are invited to actively contribute to defining the issue at stake, with the implicit assumption that by enlarging the *framing* process enough, indeterminacy can be kept under control. As Sheila Jasanoff effectively synthesizes, the European strategy to achieve "a view from nowhere," a standard of objectivity for decision-making purposes, is the so-called "view from everywhere"; that is, the inclusion of all parties involved (Jasanoff 2005). Finally, the *demarcation* model deals with the gray area, mentioned above, of the overlapping of roles between innovation and regulatory science. In particular, it is based on the idea that the institutions and the individuals that provide scientific advice have to be rigorously kept apart from the organs that make use of them. In this view, science can and has to be protected from politics and special interests.

These developments are relevant in their capacity to underline the controversial aspects of the modern ideal, but again, just as in the precautionary approach, they are designed to safeguard and preserve the privileged status of scientific knowledge and not to actually challenge its role. In Silvio Funtowicz's words: "the core activity of the modern model—the experts' (*desire for*) truth speaking to the politicians' (*need for*) power–is unchanged" (2006). In this overall view, facts and values are dangerously mixed up by our high-power capacity to modify our environment over local and global scales, and the function of normativity is to fabricate efficient defense systems to keep them apart.

TOWARD A NEW SOCIAL CONTRACT BETWEEN SCIENCE AND CITIZENS: POST-NORMAL SCIENCE AND EXTENDED PARTICIPATION

In the early nineties, around the same time as the Rio Declaration, a radically new way of conceiving the relationship between science and politics— deeply challenging the mertonian ideal of the "republic of science"—was developed by Silvio Funtowicz and Jerome Ravetz. In this framework, defined as *post-normal* science, the complexity, irreversibility and indeterminacy involved in contemporary socioenvironmental issues are fully acknowledged in all their consequences (Funtowicz and Ravetz 1993).

The starting point of their reflection is that in the majority of cases in which technoscience is involved in the normative sphere, facts are uncertain, values are in dispute, decisions are urgent and stakes are high. Moreover, the complexity implied in the entanglement between social and natural systems is multifaceted and entails that the kind of uncertainty involved is not temporary nor reducible but *radical* on the one hand, and that conflicting value positions can be supported by a plurality of legitimate and incommensurable perspectives on the other.

Indeed, a first level of complexity is intrinsic in natural systems: they are characterized by a tight coupling between different levels of organization, interacting and influencing each other through highly nonlinear dynamics, and they manifest a critical dependence from initial conditions, together with self-organizing and self-reflexive emergent properties. For these reasons, knowledge of their evolution is always incomplete and surprise is inevitable (Gallopin 2001). Jean-Pierre Dupuy defines complex systems as the ones for which the simplest model to represent them is the process itself, and the only way to predict their future is to run them. In his words: "There are no shortcuts. This is a radical uncertainty" (Dupuy 2004). But any scientific analysis has to be based on an artificially truncated system, and the boundaries that are relevant for scientists do not necessarily coincide with the ones that are significant for decision makers. This, in turn, introduces a second level of complexity, which we have previously referred to as *indeterminacy*, and which determines the coexistence of different and mutually exclusive representations of the issues at stake. Daniel Sarewitz specifically elaborates on this inherent lack of a unique perspective in terms of *excess of objectivity*: "for a given value-based position in an environmental controversy, it is often possible to compile a supporting set of scientifically legitimated facts." (Sarewitz 2004, 389). Therefore, certainty—here conceived in terms of *agreement*—cannot be achieved by looking for new scientific data and new prediction technologies alone, but on the contrary, "the value bases of disputes underlying environmental controversies must be fully articulated and adjudicated through political means *before* science can play an effective role in resolving environmental problems" (Sarewitz 2004, 385).

Finally, a third fundamental feature of the complexity involved in the interaction between technoscience, environment and society is the so-called *reflexive complexity*, given by the presence in the socioenvironmental systems of elements provided with individuality, intentionality, autonomous aims, foresight, symbolic representations and morality (Funtowicz and Ravetz 1994). The post-normal model is designed to incorporate and deal with all these levels of complexity.

A synthesis of the Funtowicz and Ravetz's framework is given in Figure 1.1.

From the one-dimensional incremental ideal of progress towards certainty, typical of the modern model and of its developments, here we step into a two-dimensional representation, where uncertainty is correlated with decision stakes. This correlation allows discriminating between three fundamentally different kinds of scientific research. In the scenario that we have so far outlined, the transition from one area to the next is determined by incrementing technoscientific power and therefore the complexity involved.

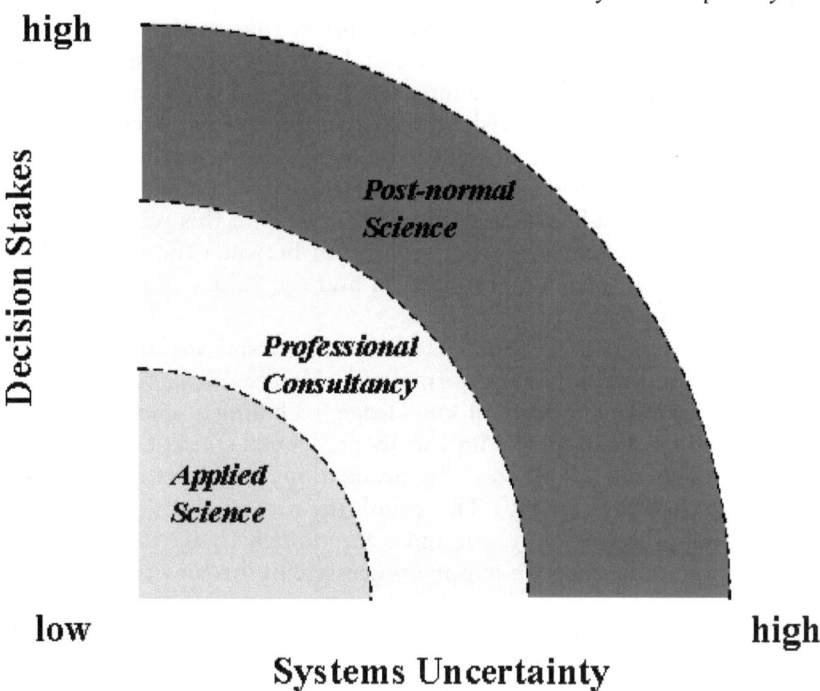

Figure 1.1 The framework of post-normal science (Funtowicz and Ravetz 1993).

Applied science is essentially laboratory science in which risks are predictable and under control. It is the condition in which the modern model has emerged and it has been easily applied.

When complexity grows and both decision stakes and uncertainty increase, we get into the area of *professional consultancy*, where specifically selected experts' advice is required to inform policy makers in order to orient the decision in the most rational and responsible way. Regulatory science and the attempts to extend the modern ideal by protecting scientific knowledge from uncertainty, indeterminacy and abuse can be ascribed to this area.

Extending technoscientific power further, we get into the *post-normal science* scenario, characterized by the paradoxical situation in which knowing more entails new levels of complexity and therefore more uncertainty, indeterminacy and ignorance. This dynamic implies the necessity of developing new methods of knowledge production and new criteria for assessing their quality (Funtowicz and Ravetz 1990).

In the modern model of applied science, relevant knowledge is *scientific* by definition, and the quality of scientific knowledge is identified with its degree of "truth," which is autonomously evaluated within the scientific community itself through the peer review procedure. The implicit assumption is that the normative sphere can and has to be kept apart from the quality assessment process.

In the developments of the modern model that we have considered, relevant knowledge is still *scientific*, and the quality assessment procedures based on peer review are integrated with new ways to evaluate the degree of *acceptability* of risk, the precautionary approach, and more generally with the degree of political relevance of scientific knowledge, in the framing and in the demarcation model. From an epistemological point of view, though, we can observe that despite this wider set of criteria, the need to achieve actual separation between the description of the relevant facts, provided by science, and the values of politics is still firmly reconfirmed.

But in the context of complex and controversial socioenvironmental issues, as expressed in the post-normal model, facts and values are *not* separable and therefore no form of knowledge, including a scientific one, can be evaluated on the basis of a univocally predefined concept of truth. New forms of public control of knowledge production then have to be elaborated (Funtowicz and Ravetz 1990). This entails the necessity, not only ethical or political, but primarily *epistemic* and *methodological*, to extend the public participation in the decision-making processes. In this new context:

> Science is considered as part of the relevant knowledge and it is included as a part of the probative evidence of a process. The ideal of rigorous scientific demonstration is thus replaced by an ideal of open public dialog. Inside the knowledge production process, citizens become both critics and creators. Their contribution has not to be defined as 'local', 'practical', 'ethical' or 'spiritual' knowledge, but it has to be considered and accepted as a plurality of rightful and coordinated perspectives with their own meaning and value structures. (Liberatore and Funtowicz 2003, 147, as cited by Tallachini 2005)

The point is not to give up scientific knowledge or to diminish its value, but to initiate a process of democratization of expert knowledge—or better, of *expertise*—by legitimizing citizens as critics through what Funtowicz and Ravetz have defined as a process of *extended peer review*. Concurrently, the *creative* role of citizens in producing relevant knowledge also has to be recognized. Including a plurality of different expertise in the decision-making processes corresponds to *expertising* the democratic procedures. The post-normal framework, also called the "extended participation" model, entails therefore a radical revision of the relationship between science and policy through a process that, on the one side, challenges the privileged role of science, and that reconsiders democracy, not only as the predominance of the majority but more properly as the unauthoritative characteristic of any language, on the other (Tallacchini 2005). This major transition is what Angela Liberatore and Silvio Funtowicz have defined as a process of "democratizing expertise and expertising democracy" (Liberatore and Funtowicz 2003).

New frameworks of interaction between science and the normative sphere have emerged beside the post-normal model, such as *Mode 2* science (Gibbons et al. 1994), sustainability science (Gallopin 2004) and the acronym SHEE—Safety, Health, Environment, Ethics—conceived to compensate the high-power innovation technosciences GRAINN (Ravetz 2006). All these models are based on the idea that quality control in producing relevant knowledge to deal with socio-environmental issues has to coincide with accountability and that the explicit involvement of citizens and stakeholders is not only a political option, but mostly an epistemic requirement. In practice, as we will see in more detail in the next chapter, this enlargement of participation can be realized through the institution of extended communities of peers, such as popular juries, focus groups, consensus conferences or through forms of spontaneous aggregations around specific issues, thanks also to the web and the information and communication technologies.

NARRATIVES AND POLICIES OF SCIENCE: FROM HUBRIS TO HUMILITY

Creating the structures of participation is a necessary but not sufficient condition to actually democratize knowledge and expertise in the policy-making processes: increasing the number of voices may not lead by itself to wiser decisions. One indeed has to reflect at the same time on how to channel collective imagination, how to stimulate public awareness, critical and creative thinking. In fact, the risk of extending participation in itself is to harden a few already predefined positions or, at the other extreme, to endlessly deconstruct the decision-making process.

In a recent report on science and governance for the European Commission (Wynne et al. 2007), the authors openly reflect on the existence of shared narratives that shape the collective imagination and determine a set of "givens" upon which policy making is implicitly based. One of the most general of these narratives is the "myth of progress", the one that associates social progress with technological advancement. Another considerably significant master narrative, correlated to the myth of progress, is the one that indicates a privileged type of knowledge in modern scientific rationality, intrinsically more valuable and effective than any other kind of cultural practice in interacting with nature (Leach and Fairhead 2003). The normative power of these narratives rests on the mechanism of selecting what counts as "evidence" and what *is not* and *needs not* to be seen. In this respect, as we will specify in Chapter 4, the use of natural and visual language plays a crucial role.

What we referred to so far as the modern model may indeed be thought of as a master narrative, with specific normative implications based on the ideals of certainty, objectivity and exhaustiveness, which was functional

Figure 1.2   Post-normal science lab.

when applied to low-power laboratory science, but has become dangerously dysfunctional in the case of contemporary high-power technoscientific research and implementation.

Sheila Jasanoff provides a specific articulation of these narratives in
her work on *technologies of humility* (Jasanoff 2003). With this term,
the author refers to the need to develop new social technologies to open
up the black box of the theoretical assumptions, the narratives, behind
the predictive, analytical and quantitative methods of innovation science, conceived to keep uncertainty under control and to reassure citizens on the technoscientific capacity to predict and solve the problems
it creates. These models, such as risk management, cost–benefit analysis
and systems simulations, are defined as *technologies of hubris* for three
main reasons.

First, they focus on the known at the expenses of the unknown, on short-
term, quantitatively manageable risks, leaving long-term indeterminate and
possibly ignored consequences—the so-called *unknown unknowns* (European Environmental Agency 2001)—in the shadow. This overstatement of
the known is based on a narrative of control that associates quantitative
technical proficiency and computation power with accuracy and complete-
ness. As we will see in Chapter 4, this narrative is articulated not only
through the aura of certainty that surrounds the symbolic language of
mathematics, but also through the aura of objectivity that is embedded in
the visual language of laboratories.

Second, the specialized language and knowledge used to elaborate and
make use of these technologies in the context of policy tend to preempt
an open discussion with *all* legitimate stakeholders. More specifically, the
normative assumptions of these analytic models are not subject to general
debate and the modern ideal of objectivity is used as a tool to obscure the

boundary work that is needed to design them. Here again the narrative of modern scientific knowledge as a privileged and neutral lens to represent and deal with reality around us plays a fundamental role in concealing the exercise of judgment and power implied in their construction.

Third, their capacity to internalize challenges that come from outside their framing assumptions is limited, as in the case of chemical toxicity assessments that continue to rest on "the demonstrably faulty assumption that people are exposed to one chemical at a time" (Jasanoff 2003, 239). This lack of epistemic plasticity is based on a narrative of exhaustiveness embedded in the modern model, which implies, as we have seen, a tendency to apply linear monocausal interpretative schemes.

Just as the "disciplines of care," defined as SHEE, should be empowered in order to balance the high-power technosciences identified as GRAINN in Ravetz's framework (Ravetz 2006), here, new *technologies of humility* have to be designed to complement the highly powerful and specialized predictive methodologies and approaches of innovation science in order "to make apparent the possibility of unforeseen consequences, to make explicit the normative that lurks within the technical, and to acknowledge from the start the need for plural viewpoints and collective learning" (Jasanoff 2003, 240). In Jasanoff's scenario, the need to extend public participation and to democratize knowledge production and its quality assurance is conveyed through four paths, which we may define as four *narratives of humility*: framing, vulnerability, distribution and learning.

Framing refers to the need to adopt systematic procedures to collectively evaluate and discuss the initial assumptions on the basis of which assessments and predictions are made. Simulation models of climate change provide an interesting example of how, once the modern narratives are made explicit, a technology of *hubris* can be reinterpreted and used in a participatory way. A shift in the epistemic *status* of simulation models was proposed by Jerome Ravetz in relation to a series of focus group experimentations on climate change conducted with more than 600 citizens around Europe (Ravetz 2003). Instead of valuing their factual content, based on their more than controversial predictive power, Ravetz elaborated on the idea of interpreting and using them as metaphors, enabling citizens, including the software engineering experts and modelers, to express their presuppositions, as well as their fears and hopes, about the issue at stake. The result was a significant enlargement of people's imagination on the climate change issue and on the role of individuals as actors of the problem and of its possible solutions.

Vulnerability points to the need to open up to public discussion the ways in which individuals and populations are classified into different groups according to their exposure to risks. The modern narrative of rational scientific knowledge entails a classification mostly based on biomedical data and monocausal thinking, such as genetic predisposition, gender and age, disregarding the socioenvironmental factors such as history, geographical location and degree of connectedness with the natural and social environment.

A typical case of such a reductive strategy, which has significant normative implications, is the classification of people according to the risk of dying of malaria, mostly based on age and gender, and disregarding socioenvironmental factors, such as primary access to food, clean water and free medical treatment. Again, extending the public participation on the issue of vulnerability opens for discussion and deconstruction the narrative of control and privilege embedded in the Western technoscientific way of thinking.

Distribution refers to the fact that ethical and political debates about the impact of innovation sciences are too often confined to the limited area of risk, leaving aside the issue of equity and the social and economical realignments that they imply. The dominant discourse on GMO, for example, is founded on the narratives of *fear* on the one side—of possible human health and irreversible environmental damages—and of *urgency* to tackle and solve the global issue of starvation in southern countries on the other side. The social and economical implications of the implementation of GMO technologies—more specifically, the distributive impact of the patenting mechanism that transforms public knowledge and natural resources into private, corporate know-how—are rarely opened up for public discussion (Shiva 1997; Berlan 2001). To make a narrative of distribution explicit, and to collectively share it, would again have fundamental normative consequences, in this as in many other cases of GRAINN technosciences.

Finally, a participatory approach to *learning* in science for policy-making implies an open discussion on the assumptions about what is to be learned and how. As we have seen, high-power technoscience is experimented directly on the socioenvironmental systems of the planet, and learning from experience is quite a different procedure than learning from experiment. In particular, as Jasanoff points out, experience is polysemic, and it is therefore intrinsically interpretable in many different ways through a multiplicity of narratives. Monocausal explanations and the narratives of unintended consequences are but attempts to extend the modern ideal of control and certainty out in the world of complexity and controversy, by exempting both technoscience and the normative sphere of ethics, politics and the law from accountability in socioenvironmental controversies and in "normal accidents." Just like historians, scientists, policy-makers and civil society then have to engage in a process of reflection about past events and future possibilities through the development of systemic thinking and mostly through the retrieval of individual and collective creative potential. Dominant narratives that inform learning as well as knowing and imaging processes have to be uncovered and complemented with a plurality of approaches, methodologies, languages and interpretation schemes democratically elaborated and shared. Education becomes, in this view, a fundamental tool of empowerment and democratization of not only of knowledge, but also, most importantly, of imagination and creativity.

BIBLIOGRAPHY

Beck, U. 1992. *Risk society: towards a new modernity*. London: Sage.

Berlan, J.P. ed. 2001. *La guerre au vivant: organismes génétiquement modifiés et autres mystifications scientifiques*. Marseille, France: Édition Agone.

Commission of the European Communities. 2000. *Communication from the Commission on the Precautionary Principle*. Brussels, Belgium.

Dupuy, J.P. 2004. "Complexity and Uncertainty a Prudential Approach to Nanotechnology," in European Commission, *A Preliminary Risk Analysis on the Basis of a Workshop Organized by the Health and Consumer Protection*, Directorate General of the European Commission.

European Environmental Agency. 2001. *Late Lessons from Early Warnings: The Precautionary Principle 1896–2000*, http://www.eea.eu.int

Fairhead, J., and M. Leach. 2003. *Science society and power: environmental knowledge and policy in West Africa and the Caribbean*. Cambridge: Cambridge Univ. Press.

Funtowiz, S., and J. Ravetz. 1990. *Uncertainty and quality in science for policy*. Dordrecht, The Netherlands: Kluwer Academics Publishers.

———. 1993. Science for the post-normal age. *Futures* 31(7): 735–55.

———. 1994. Emergent complex systems. *Futures* 26(6): 568–82.

———. 1999. Post normal science: an insight now maturing. *Futures* 31(7): 641–6.

Funtowicz, S. 2006. *Why knowledge assessment*. In *Interfaces between science and society*, ed. Â. Guimarães Pereira, S. Guedes Vaz, and S. Tognetti, 138–45. Sheffield, UK: Greenleaf.

———. 2007. Modelli di scienza e politica. In *Biotecnocrazia: informazione scientifica, agricoltura, decisione politica*, ed. C. Modonesi, G. Tamino, and I. Verga, 55–71. Milano: Jaca Book.

Gallopin, G.C., Funtowicz, S., O 'Connor, M., and Ravetz, J. 2001. Science for the 21st century: from social contract to the scientific core. *International Journal of Social Science* 168: 219–29.

Gallopin, G. 2004. Sustainable development: epistemological challenger to science and technology, ECLAC, Santiago del Chile.

Gibbons, M.H., Nowotny, C., Limoges, C., Schwatzman, S., Scott, P., and Trow, M. 1994. *The new production of knowledge: the dynamics of science and research in contemporary society*. London: Sage.

Jasanoff, S. 1990. *The fifth branch: science advisers and policymakers*. Cambridge MA: Harvard University Press.

———. 1995. *Science at the bar*. Cambridge, MA: Harvard Univ. Press.

———. 2003. Technologies of humility: citizen participation in governing science. *Minerva* 41(3): 223–44.

———. 2005. *Designs on nature*. Princeton, NJ: Princeton Univ. Press.

Jonas, H. 1985. *The imperative of responsibility. In search of ethics for the technological age*. Chicago: Univ. of Chicago Press.

Liberatore, A., and S. Funtowicz. 2003. "Democratising" expertise, "expertising" democracy: what does this mean, and why bother. *Science and Public Policy* 30(3): 146–50.

Merton, L.K. 1968. *Science and democratic social structure in social theory and social structure*. New York: Free Press.

Perrow, C. 1984. *Normal accidents: living with high risks technologies*. New York: Basic Books.

Ravetz, J. 2006. *The no nonsense guide to science*. Oxford: The New Internationalist.

———. 2003. Models as metaphors. In *Public Participation in Sustainability Science: A Handbook*, ed. B. Kasemir, J. Jäger, C.C. Jaeger, and M.T. Gardner, 62–77. Cambridge: Cambridge Univ. Press.

Sarewitz, D. 2004. How science makes environmental controversies worse. *Environmental Science and Policy* 7: 385–403.

Shrader-Frechette, K.S. 1996. Methodological rules for four classes of scientific uncertainty. In *Scientific Uncertainty and Environmental Problem Solving*, ed. J. Lemons, 12–40. Oxford: Blackwell.

Singer, P. 2002. *One world: the ethics of globalization*. New Haven, CT: Yale Univ. Press.

Shiva, V. 1997. *Biopiracy: the plunder of nature and knowledge*. Cambridge, MA: South End Press.

Smith, R., and B. Wynne, eds. 1989. *Expert evidence: interpreting science in the law*. London: Routledge.

Tallacchini, M.C. 2005. Before and beyond the precautionary principle: epistemology of uncertainty in science and law. *Toxicology and Applied Pharmacology* 207: 645–51.

Wildavsky, A. 1979. *Speaking truth to power*. Boston: Little Brown and Co.

Wynne, B., Felt, U., Callon, M., Gonçalves, M.E., Jasanoff, S., Jepsen, M., Joly, P.B., Konopasek, Z., May, S., Naeubauer, C., Arie, R., Siune, K., Stirlig, A., and Tallacchini, M.C. 2007. *Taking knowledge society seriously*. European Commission: EUR 22700 EN.

2 Post-Normal Relationships between Science and Society
Implications for Public Engagement

Ângela Guimarães Pereira

Science and technology are present in all of the narratives that modern societies weave about the world, as essential threads in the tapestry of social reality.

(Jasanoff, 2005)

EDITORS' INTRODUCTORY NOTES

The epistemological changes that characterize modern technoscience, as described in Chapter 1, along with the growing concentration of economic power and energy, have important implications for the involvement of the public in decision-making processes concerning socioenvironmental issues. In this view, the epistemic framework of post-normal science challenges the model of Public Understanding of Science (PUS), and the thesis of the public's inability to act on scientific issues, supporting instead the launch of various efforts for democratizing science. Such an extension of participation calls for a reflection on the nature of the knowledge that is being produced and on what the criteria and processes for ensuring quality might be. Chapter 2 deals with these questions through an exploration of different models of participation and co-construction of knowledge, which refer to different perspectives on the relationships between science and society. In particular, if exclusive reliance on scientific knowledge validated by the academic community is no longer tenable, then quality in decision making is linked to the extent to which multiple subjects can be involved and participate in a process of dialog: an extended peer community that is able to balance previous power differences between science and society, and also to enable co-production of civic science.

This chapter will offer a number of examples of contexts and places for engagement, with a final reflection on the changes in the ways of learning that are required by the ongoing transformation of the relationship between the public and the scientific community.

SCIENTIFIC KNOWLEDGE AS A POWERFUL
MEANS OF REPRESENTING REALITY

One of the most important challenges I have had recently has been in trying to explain to my four-year-old son some of his curiosities about the world. I now have to explain everything I have (unconsciously) taken for granted. Even if fantasy helps a great deal, most of the time my tales and metaphors are based on science. This is hardly surprising given my engineering background, and also probably because I, too, as a child, was not exposed to other types of narratives, even although I lived in Africa.

Every day we *translate*[1] scientific knowledge, or are exposed to others' translations, in our metaphorical personal world, using our own vocabulary, our referential walks of life, and apply it in our context of significance. Doctors translate illness information to their patients; parents respond to their children's questions with science tales; the press, movies, radio or TV shows convey scientific stories to their audiences; and arts often express scientific views. In regulation, as in the European environmental impact assessment (CEC 1997) and the water framework directives (CEC 2000), a nontechnical summary of the specialist reports is requested to be available to the nonspecialist public. Likewise, among specialists, exercises of *translation* are done, as, for instance, when the European Food Safety Agency writes opinions to the European Commission's Directorate General Health and Consumption (DG SANCO).

Moreover, the producers of science, generally denominated as scientists, are often seen as privileged with a power that derives from their proximity to the truth. Hence, it makes sense that their narratives become tangible and accessible to all! The US educator John Dewey claimed that young people should be inculcated with a "scientific attitude" that would help them approach the issues and problems of everyday life in a rational and logical fashion (in Dewey [1934], quoted in Gregory and Miller [2000]).

From a historical perspective, science as we know it appeared in the seventeenth century and was accompanied from day one by science popularization and science fiction. One of the first science popularization books was Galileo's "Dialogue," published in Italian in February 1632. Athanasius Kircher used the then-recent optical scientific discoveries to design spectacular public shows in Rome around 1640 and created one of the first science museums. The science-fiction novel, "Les états et empires de la lune et du soleil" by Savinien Cyrano de Bergerac (a physicist) was published in 1648.

Also, quite early public involvement in science was sought as a means of legitimation. For example, part of the mission of the Royal Society of London, which was founded in 1660, was public demonstrations of new science as part of its validation process[2]. Historians of science have pointed out the fact that prior to the days of peer-reviewed journals and elite science societies, the latest facts and theories about science were regularly discussed in public (Miller 2001). In a way, what we understand today as "peer review" has killed extended debate and promoted elitism and closure of the scientific community.

The national history or tradition of a country has greatly influenced the actions taken to promote research and technology development (RTD) culture and public involvement in science in that country (Miller et al. 2002; Jasanoff 2005). Some historical crisis of science–society relationships has set the stage for present day attitudes, science communication styles having taken different routes in different countries. In the remainder of this chapter, I will look at the ways in which the public's involvement in policy-relevant science has been conceptualized, especially in Europe. I will also look at how, in practice, researchers view engagement of the public in the development of science. I will look into the challenges of democratizing science and how extended frameworks, such as knowledge co-production (Jasanoff 1994) and the post-normal science framework (Funtowicz and Ravetz 1990, 1991, 1992) in which the concept of extended peer review and extended peer communities have developed, can contribute to its *operationalization*. Finally, I will reflect on the implications for education and learning processes of the introduction of these extended processes for knowledge creation.

MODELS OF PUBLIC INVOLVEMENT IN SCIENCE

In this section, I concentrate on the publics' involvement in policy-relevant science in the last decades. The types of processes I am considering here relate to situations where science is used to inform a debate about a specific policy issue. In particular, I will adopt the three main phases of "public involvement" in science identified by Wilsdon and Willis (2004): a) public understanding of science; b) from "deficit to dialog" and c) engagement upstream. They also find correspondence with Callon et al. (2001) and others (see, for instance, Felt 2002) models of science and public relations.

a. PUS

The PUS (or PUST if the word *technology* is added, or yet PUSH if the humanities are considered, as in the German approaches to the issue) developments since the mid-1980s arose on the assumption that the lack of public support for science and innovation was solely due to lack of understanding or "scientific illiteracy" (Wynne et al. 2007), which was also deemed to jeopardize modern democracies. This concept corresponds to Callon's (2001) model of science and public relations called the "public education model."

The response[3] *from scientists* to growing levels of public detachment and mistrust was to embark on a mission to inform, a one-way transmission of knowledge from science to a public imagined as passive and lacking in information. It is, in practice, a pedagogical approach that relies on a "deficit model" of the public as ignorant and science as unchanging and universally comprehensible and, therefore, tries to increase the scientific knowledge of citizens. Some have even elaborated practical protocols with sound principles to engage in such an endeavor like Gregory and Miller (2000)—although many aspects of the latter are much beyond the PUS paradigm.

Many have argued against this paradigm because of its ambiguities, like Duran et al. (1996), who discussed the very expression: what is the "public," what is "understanding" and what is "science"?, and Jasanoff (2005), who argues that the "u" of understanding in PUS implies the interest of the public in scholarly issues, potentially being, in this case, a source of cross-cultural variance: failure to understand science becomes a meaningful dimension of difference among individuals and communities. Gross (1994), in his work about the rhetoric of PUS, outlines how the deficit model incorporates the metaphors of scientific *sufficiency* and public *deficiency*: "its practitioners do not try to persuade, they assume that the public is already persuaded of the value of science" and "they do not try to build trust; they assume that the public is already trusting"; hence, this model implies a passive public, which "requires a rhetoric that acts to accommodate the facts and methods of science to the publics' limited experience and cognitive capacities" (Gross 1994, 6).

In much of the research that this concept has generated, there is little evidence that public ignorance of scientific facts correlates in any meaningful ways with collective responses to science and technology; biotechnology being simply one domain for which this observation holds true (Gross 1994; Evans and Durant 1995; Wynne 1995; Miller 2001; Jasanoff 2005).

b. From Deficit to Dialog

The UK BSE[4] "scandal" of the mid-1980s to mid-1990s is often cited as pivotal in the change of direction in the relations between science and policy making. A key moment was the publication of the 2000 House of Lords report on *Science and Society*, followed a year later by the European Commission's *Science and Society Action Plan* (European Commission 2002), as well as the European Union (EU) 5th Framework research program's "Raising Awareness of Science and Technology" activity of the late 1990s. Partly as a result of PUS failings, public perception of science worsened throughout the 1990s, and a new language of "science & society" towards dialog engagement emerged. This phase corresponds to Callon's (2001) model of science and public relations as a "public dialog and participation model."

"The GM Nation?"—see Box 2.1.—is an example of this phase in which the UK government sponsored a debate on genetically modified crops with the intention of having a wide-ranging and effective public debate, going beyond the "often polarised views in order to find out what people really

---

Box 2.1   The GM Nation?

The public debate in the United Kingdom that took place during 2003 about genetically modified organisms was a unique experiment in public engagement. It involved a total of 675 public debate meetings, namely:

- 9 Foundation workshops with members of the public

- 6 National and regional conferences; small county and local-level meetings

- Focus groups (*narrow but deep* meetings)

- Material: CD-ROM, brochures incl. questionnaire

- A website where people could post comments and accompany progress of debate.

(http://www.gmnation.org.uk/)

---

think about GM" (Gaskel et al. 2003). It is important to note that this debate springs from a recommendation of the *Agricultural and Environment Biotechnology Commission*, which is an independent body that advises the UK government on biotechnology issues and their impact on agriculture and the environment. This debate should "establish the full spectrum of the publics views on GM and possible commercialisation of GM crops, and any conditions it might want to impose on this" (AEBC 2003). It is remarkable that the GM Nation was a governmental operation: as Jasanoff (2005) notes, in biotech times, upstream efforts to identify risks and explore ethical dilemmas were led by the science community itself. Indeed, many initiatives of public involvement in controversial issues depart from the academics or nongovernmental organizations and fewer from governmental institutions.

Horlick Jones et al. (2004) and Rowe et al. (2005) made the point that the public was concerned about lack of transparency, especially in terms of how results would be used, as well as the actual influence of the views expressed in the whole policy process, from events that had not been held early enough to influence public policy. Moreover, Wilsdon and Willis (2004) highlight that the publics involved believed that there was a genuinely open outcome at stake and that everyone believed that the government had already determined its preferred outcomes. As the government did not ascertain its position, a deep public sense of government dishonesty developed and was aggravated by the government being the sponsor of this initiative.

c. Moving Engagement Upstream

In this (current?) phase, the science community is (supposedly) embracing dialog and engagement (recognizing that many controversies had made it a nonnegotiable clause of their "license to operate"). A new term has entered the lexicon of public engagement: there has been a surge of interest in moving engagement "upstream" (e.g., The Royal Society's nanotechnology report in 2004[5]: *constructive and proactive debate about the future of nanotechnologies to be undertaken now . . .*); it focuses on establishing a two-way dialog between citizens and other actors on science and technology challenges facing society.

This phase corresponds to what Callon (2001) describes as the model "public co-production of knowledge," as far as science and public relations are concerned.

But how do researchers view public engagement in policy-relevant science at present?

MOTIVATIONS AND PLACES FOR PUBLIC INVOLVEMENT IN SCIENCE

Researchers Views about Public Engagement in Science

In this section, contemporary views of researchers with regards to public engagement in science will be reviewed. The observations herein are based on a report from the UK Royal Society published in August 2006 (Royal Society 2006) regarding the factors affecting science communication undertaken by scientists and engineers. This report is based on a survey done with UK researchers, which makes it specific to the United Kingdom in context, but (based on experiences reported elsewhere with other researcher communities) many of the observations of interest in this chapter apply to many other contexts.

In the Royal Society report of August 2006, scientists were asked to define, in their own terms, what *engaging with the nonspecialist public meant to them* (the percentage that considered the meanings are found in parentheses below):

- To explain and promote PUS (34%)
- Highlighting implications, relevance and value of science (15%)
- Giving a public lecture (13%)
- Listening to and understanding the public (13%)

Moreover, respondents considered that:

- The most important reason to engage the nonspecialist public is to ensure the public is better informed about science and technology (35%)

- The least important reason was to engage the nonspecialist public to contribute to ethical discussions about science (5%)

Yet, 69% of the respondents agreed with the statement that it is a moral duty of scientists to engage with the nonspecialist public about the social and ethical implications of their research.

The interviewees saw the following as barriers to science communication and involvement with the public:

- The need to spend more time on research (64%)
- Time taken away from research (29%)
- Scientists who engage with the public are less well regarded by other scientists (20%)
- Several researchers highlighted that public-engagement activity was seen by peers as bad for their career, "done by those who are not good enough" for an academic career, with science communication being regarded as an altruistic activity and not bringing significant funding to universities.

Reflections

As I said earlier, cultural and historical events in science and public relations shape current situations in different countries. Yet, 73% of the researchers that responded to the UK Royal Society survey had no training in communication. So one might think that lack of skills and current understanding of what a "normal" research activity is play a big part in the ways the public engagement in science is currently shaped in many countries, along with the historical context. Science communication and engagement with the public, as viewed by researchers, confirms that the great majority of respondents still adopt the one-way information paradigm. Callon's (2001) model of knowledge co-production, where citizens and concerned groups get actively involved in the process of knowledge production, is still far-off.

Several normative and regulatory documents coordinated by the European Commission, such as the report on the "democratising of expertise" (European Commission 2001) and the report on "taking knowledge society seriously" (Wynne et al. 2007) outlined the need for, and the conditions to, improving interactions between expertise, policy making and public debate, as well as new regimes of public engagement. Different motivations can be found to foster a "deeper" type of engagement that goes beyond the still entrenched paradigm that sustains the current interactions between expertise, policy making and public debate. These can be due to the inevitable, more inclusive ways in which democracies tend to develop, or because they legitimate a specific process; for some, because such interactions are a unique way to change innovation developments policy making or yet a reflexive and accountable review of current knowledge production activities (Wynne et al. 2007). While this surely has to be embedded in an extended model of governance of science for policy making, where the public is viewed as possessing relevant resources that may be relevant in a knowledge co-production (Jasanoff 1994) process, the operational framework is varied and still experimental in many cases, lacking, above all, the institutional link.

FROM PUS TO "CO-PRODUCED FACTS"

In this chapter, I am interested in situations where public involvement is pertinent in policy-relevant science. Therefore, I start to look at models of the relations between science and policy and see where in such models a more extended relation of the public with the science production and deployment could find its space. Subsequently I look at the notions of co-production, civic epistemologies and alternative contextual models that address some of the pitfalls of the PUS model and practice.

Models of Science and Policy: From Expert Demonstration to Post-Normal Science

Funtowicz (2006) identified several conceptual models of the relation between science and decision making in policy processes. Funtowicz traces their evolution through a deepening appreciation of the process of the use of science in policy. Starting from the «modern model» of perfection and perfectibility—which represents a classic "technocratic" vision where there are no limits to the progress of humans' control over their environment, and no limits to the material and moral progress of mankind—Funtowicz offers an evolutionary perspective on the governance of science in policy making:

- *Precautionary model* (with uncertain and inconclusive information): arises from discovering that the scientific facts are neither fully certain in themselves nor conclusive for policy; therefore, an extra, normative element is introduced in policy decisions: *precaution*, which both protects and legitimizes decisions.
- *Framing model* (arbitrariness of choice and possible misuse): arises from the recognition that, in the absence of conclusive facts, scientific information becomes one among many inputs to a policy process, functioning as evidence in the arguments. Stakeholders' perspectives and values become relevant, and even the choice of the scientific discipline to which the "problem" belongs becomes a prior policy decision, part of the debate among those affected by the relevant issue.
- *Demarcation model* (possibility of abuse of science): arises because the scientific information and advice that are used in the policy process are created by people working in institutions with their own agendas. It recognizes that the "scientific" information and advice cannot be

guaranteed to be objective and neutral. In this sense, science can be abused when used as evidence in the policy process. A clear demarcation between the institutions (and individuals) who provide the science and those where it is used is advocated as a means of protecting science from the political interference that would threaten its integrity.

- *Extended participation:* acknowledges the difficulties of defending a monopoly of accredited expertise for the provision of scientific information and advice. "Science" (understood as the activity of technical experts) is included as one part of the "relevant knowledge" is brought in as evidence to a process. The ideal of rigorous scientific demonstration is replaced by that of open public dialog. Citizens become both critics and creators in the knowledge-production process as part of an extended peer community.

It is argued that it is within the "extended participation" model of science and policy that new relations between science and the public are usefully operated, underpinning phase 3 of public engagement in knowledge production. The "extended participation" model is both framed and operationalized within the post-normal science framework and the concept of *extended peer review*, which we address below.

Post-Normal Science and Extended Peer Review

The insight leading to Post-Normal Science is that in the sorts of issue-driven science relating to environmental debates, typically facts are uncertain, values in dispute, stakes high, and decisions urgent [. . .] In post-normal conditions, such products the goal of achievement of truth or at least of factual knowledge may be a luxury, indeed an irrelevance. Here, the guiding principle is a more robust one, that of quality. (Funtowicz and Ravetz 1990b)

As reported in Chapter 1, in the diagram of post-normal science (reported on page 19 of this book), Funtowicz and Ravetz (1985; 1990a; 1992) represented how different types of problem-solving strategies and practice correspond to different sorts of uncertainty (namely technical, methodological and epistemological), as well as how they relate to the world of policy: *decision stakes* included "costs, benefits, and commitments of any kind by the parties involved."

The "post-normal science" framework (Funtowicz and Ravetz 1990b; 1992; 1993; http://www.nusap.net) and its guiding principle—quality—requires the participation of an "extended peer community" (Funtowicz and Ravetz 1990b) engaged in dialog and the resolution of the issues at stake. An "extended peer community"—see Box 2.2.—consists not merely of persons with some form of institutional accreditation, but rather all those with a desire to participate in "extended peer review" processes for

---

Box 2.2   "Extended Peer Communities."

"Extended Peer Communities" are increasing in number, organized for different purposes and involved in different ways and at different steps in policy-making processes. They are called "citizens juries," "focus groups," "consensus conferences," etc., with correspondingly varied powers. They may be engaged through regulated participatory processes, or they may be the initiative of specific interests or even develop into formal settings, resulting from social mobilization. But they all have one important element in common: they assess the quality of policy proposals, including a scientific element, on the basis of the science they master combined with their knowledge of the ways of the world. The contribution of relevant social actors, in this case, is not merely a matter of broader democratic participation, and their verdicts all have some degree of moral force and, hence, political influence (Funtowicz 2001).

---

the resolution of the issue (Funtowicz and Ravetz 1990b). Their relevance and influence will depend on the context in which they operate, the eventual aim of the involvement and the flexibility of, or instrumental use by, institutional "ears" that could benefit from their inputs.

The assessment of the *quality* of the knowledge inputs to policy issues by those "extended communities" is in many ways different from the review processes of research science, professional practice or industrial development (Funtowicz 2001). Each of those has its established means of quality assurance for the products of the work, be they peer review, professional associations or the market. However, what Funtowicz and Ravetz argued is that, for new controversial problems, the maintenance of quality depends on open dialog between all those affected.

The aim of knowledge quality assurance by "extended peer review" is precisely to open processes and products of policy-relevant science to those who can legitimately verify its relevance, fitness for purpose and applicability in societal contexts, contributing with extended insights and knowledge: "extended facts." These may include craft wisdom and community knowledge of places and their histories, as well as anecdotal evidence, neighborhood surveys, investigative journalism and leaked documents (Funtowicz 2001). Extended peer review faces several challenges, such as, for example, resistance and closure of institutional or established practice in research and policy, different conceptual and operational framings and knowledge representations and mediation (Guimarães Pereira and Funtowicz 2005).

Knowledge Co-production, "Civic Epistemologies" and Contextual approaches

This section looks at the concepts of "knowledge co-production" and "civic epistemologies" as proposed by Jasanoff (1996, 2004, 2005).

"*Co-production* is short hand for the proposition that the ways in which we know and represent the world (both nature and society) are inseparable from the ways in which we choose to live in it" (Jasanoff 2004, 3).

In this framework, scientific endeavor is not to be understood as "a simple reflection of the truth about nature nor an epiphenomenon of social and political interests" (Jasanoff 2004, 3). Scientific knowledge "both embeds and is embedded in social practices identities, norms, conventions, discourses, instruments and institutions—in short all building blocks of what we term the social" (2004, 3). Hence, scientific knowledge is not independent of political contexts, but *co-produced* by scientists and the society within which they are embedded (Jasanoff 1996).

As with a post-normal science framework, the co-production framework attempts to interpret and account "for complex phenomena so as to avoid the strategic deletions and omissions of most other approaches in the social sciences" (Jasanoff 1996). Hence, this framework conceptualizes the scientific endeavor as intrinsically embedded in the context where it develops and so scientific knowledge as being context-dependent, a notion that is further developed in Giampietro (2003), for whom scientific narratives and its framings are dependent on "who" initiates the process, with which purpose, perspectives and values.

The ideas of co-production of knowledge liaise with the motivations and justifications for the ways in which the publics hold, develop, represent, communicate or express and deploy knowledge. Jasanoff (2005) offered the term "civic epistemology" to refer to the culturally specific, historically and politically grounded public-knowledge ways; i.e., what she calls the "institutionalised practices" by which society members "test and deploy knowledge claims used as a basis for making collective choices" (255). Through this concept, one moves away from "*a priori* assumptions about what the publics should know or understand of science." "Moreover, it challenges the assumption that the credibility of science in contemporary political life can be taken for granted when, in fact it is a subject that needs explanation (Jasanoff *idem*), i.e. how science claims become authoritative and through which ways science inputs become legitimate in policy settings needs to be addressed.

The notion of "civic epistemologies" and the "extended peer review" earlier described explicitly recognize that the public at large shares certain values, experiences and ways of testing and handling knowledge claims. Hence, such acknowledgement requires the consideration of a relationship between the science and the public to be necessarily conceptualized on a basis that is different from what was offered to support the PUS movement.

Gross (1994) suggests a "contextual model" as the counterpart to the "deficit model" of PUS. In the "contextual model," he argues, scholars do not have a methodological security, as practitioners of PUS—who often use surveys and statistical analysis to justify their approaches—claim to have. The author suggests that case studies should, in this case, be considered as a social scientific method. In the "contextual model," interaction between science and its public is the basic metaphor; hence, it depicts communication as a two-way flow between science and its public. In this model it is not assumed that the public is already trusting, implying also a very active public. This model implies the rhetoric of reconstruction, where public "understanding" is the *joint creation of scientific and local knowledge; the genre is deliberative*. Ethical and political concerns are always relevant.

From PUS to "Co-Produced Facts"

Jasanoff's "co-production" framework and "civic epistemologies" help us to realize that the scientific endeavor has indeed been a collective one, for it results from an implicit or explicit interaction of "science makers" and the contexts in which they operate. This is an important recognition, since those in charge of scientific developments often assume positions of "independence," objectivity, value and passion-free and neutrality, which become arguable in the light of the thorough acknowledgment of the dependencies of science in its context.

Post-normal science and its accompanying concepts of "quality assurance" through "extended peer review" by "extended peer communities" with their "extended facts," as well as manifestations of more inclusive participation of the public in science production, such as those of the European science democratization (CEC 2001), proposals seem to be—in a way—a natural path to follow. In a previous section we argued that it is within the "extended participation" model of science and policy, as proposed by Funtowiz (2006), that new relations between science and the public may be usefully operated because it encourages (and justifies) public engagement in policy-relevant science developments. The "extended peer communities" that this model alludes to are the key concepts in this chapter that justify public engagement in science, i.e., the acknowledgment that publics are not passive recipients of knowledge inputs provided by specialists, but are able to engage in "co-production of facts."

As we have seen in the previous sections, several authors, while offering different models of the relation between the public and science production, have demonstrated the pitfalls of the "deficit model" as the means of interfacing science and society, e.g., Gross's (1994) "contextual model" as presented earlier. In contrast with the "deficit model," contextual-like models have no prior assumptions about public knowledge and view science–society relationships as a cooperative endeavor, leading to "co-produced facts." The latter could be defined as the knowledge produced as a result of partnerships among those who have relevant knowledge that helps taming a specific *problematique*, whether it is a result of shared framing, data collection, shared scope, analysis or other forms of knowledge creation.

The concepts presented earlier are all important to understand what questions the PUS paradigm does not address, making it irrelevant for collective action. For instance:

- How knowledge is produced, communicated, legitimized and deployed for collective action;
- What are the context dependencies to be considered in a knowledge mediation process;
- What form *interfaces* between science and society might have when the objective is one of debate instead of education.

The ways PUS practice has been operating, and its more urgent question of what the public knows about science, is relevant for the sole purpose of deciding what requirements are needed to implement a strategy of knowledge mediation when dialogs with the public are desirable. Moving beyond the *reductionist* relationship between science and society that PUS implies requires rethinking the "safe spaces" for knowledge exchange in public debate, with methodological implications for implementing those debates including the knowledge communication. The next section will reflect on this.

PUBLIC ENGAGEMENT IN SCIENCE AND TECHNOLOGY

> Public engagement today can directly affect research. It has gone be-
> yond debating controversial social impacts of applied science and tech-
> nology (. . .). It now delves into research methods that are unique to the
> laboratory, such as somatic-cell nuclear transfer and hybrids. (Taylor
> 2007, 163)

A whole issue on science and policy published by *Nature* in September 2007 heralds public engagement in science and technology as an inevitable and well-grounded activity for research operation. There is a whole body of literature, mainly based on case studies, reflecting on the challenges, opportunities and motivations for the public to be engaged in the science production processes, especially when dealing with policy-relevant science. "Public engagement" is viewed with enthusiasm to implement more inclusive governance styles. The implementation of participatory activities within policy- and decision making has evolved, nowadays being the subject of regulation and becoming more formalized within existing regulatory frameworks (see, e.g., De Marchi et al. 2001). Still, oftentimes, in public engagement, the public is involved in some limited manner in the practices of policy making, in contrast with the predominant model of representative democracy, where the public is "involved" solely by voting and election mechanisms (Guimarães Pereira et al. 2005; Rowe et al. 2005). The question now seems not to be whether public involvement should occur— there is a great deal of normative argumentation for doing it. The questions now seem to be about how it should occur, at which point of the process, its impacts and in what fields is it legitimate and relevant (Joly and Rip 2007; Taylor 2007). Notwithstanding the more accommodating tendency, effective public engagement exercises are full of difficulties of theoretical (what is effectiveness in this context?), practical (how do we assess that?) and political (how can this be done in often contested terrains?) natures (Rowe et al. 2005). Elsewhere (see, e.g., De Marchi et al. 1998), a number of principles essential to ensure an effective involvement of "extended peer communities" were described. A first principle, "knowledge-sharing," refers to the necessity of recognizing and appreciating the different types of knowledge that different agents can bring into a dialog. For example, citizens exposed to a certain risk are not a tabula rasa. They derive much relevant knowledge from everyday experience, dealing with real world problems. Acknowledgment of a community's "resources" refers to all the available talents, expertise, connections, etc. of different community members, which also include social and communicational skills, as well as access to extended networks. A policy dialog facilitates the elicitation of such resources that, once discovered, can be enhanced and used in a social learning process.

Many argue that changes are necessary in the institution and culture of science to one that supports more participatory and deliberative research designs (Cortner and Moote 1999; Bellamy and Dale 2000; Funtowicz et al. 2000; Wondolleck and Yaffee 2000; Pound et al. 2003; Bellamy et al., 2004; Burgess and Clark 2006; Wynne et al. 2007). Joly and Rip (2007) sustain that public engagement in science and technology is thriving, especially in the United Kingdom, suggesting that, in a number of case studies, such involvement has been fruitful for scientists and members of the public.

In order to implement participatory activities, practitioners often assume that citizenship is given. Writers about these issues often focus on how best to involve people in policy and decision processes and less on what it means to be a citizen and how its varied attributes express in public engagement (Jasanoff 2004); in other words, also focusing on the bases for inclusion and exclusion instead of the purposes served by wider inclusion.

We would argue that these issues are not disconnected; the loose link that these processes often have with their potential institutional "ears" is, at present, the most important bottleneck of these processes. So, when examining the methodological aspects of public participation implementation, one should look at the reasons why the civic and the political worlds do not connect, examining expectations from those involved and the meanings of the involvement. Wynne et al. (2007) distinguish between invited and uninvited participation, explicit and implicit forms of participation and the private and public arenas (e.g., media debates)—see Box 2.3.—for the meanings. These are contextual aspects of involvement of the publics. They frame the ways in which such processes are conducted, both conceptually and methodologically speaking.

Although those who advocate the democratization of science and more inclusive science and governance approaches certainly have a reason to

---

Box 2.3   Contextual aspects in which public involvement may occur.

- **Invited/uninvited** participation concerns the legitimacy of societal voices and how they are determined in relation to the framework where they operate;

- **Explicit or implicit** participation tries to capture the fact that the public can be absent presences in the sense that proxies might be used to integrate the publics' views in policy making processes, such as survey results or similar tools, or by more informal processes.

- **Private and public arenas** set also the context in which public engagement is framed; one around a vision of citizens as individuals, and the other around the public as represented through *stakeholders*[6].

Based on Wynne et al. (2007).

---

cheer the evolution it has taken over the last decades, we are still far from an extended model of science and policy.

Places of Engagement in Science and Technology

As we said earlier, public involvement methodologies are now beyond experimentation. There is an immense body of literature on participatory methodologies (see, for instance, Morgan 1998; Glicken 1999; IAP2 2000; Frewer et al. 2001; Van Asselt et al. 2001; Peals 2003; Slocum 2003; Involve 2005) that aims at *operationalizing* public involvement, discussing their appropriateness and conditions for implementation. Having said that, there is space to discuss whether those methods, taken from social research, are the appropriate interfaces between the public and institutions involved in the science–policy cycles. Although the argument for lack of connectedness between participatory initiatives and institutions is grounded on stands of political and cultural nature, there is no evidence that the methodologies chosen are not playing a hindering role as well. For instance, a recurrent argumentation about using qualitative methodologies in social research is the issue of *representativeness*, which links closely with the methodologies chosen and social–political context.

In the European Union (EU), two-thirds of Member State Governments now either have, or are developing, mechanisms of involving the general public in issues concerning scientific and technological developments (Austria, Denmark, Finland, France, Greece, Germany, Ireland, The Netherlands, Sweden and United Kingdom [Miller et al. 2002]). Denmark currently has the most extensive toolkit of participatory instruments, organized through the *Danish Teknologirådet* (http://www.tekno.dk/)—the Danish Board of Technology. Denmark was also the EU country that pioneered Scientific Ethical Committees to approve research procedures, such as medical trials. In fact, this Board, which informs parliamentary deliberations and decisions with well-crafted forms of direct elicitation of relevant public concerns, meanings and attitudes, is one of the exceptions to the otherwise overall realization that many participatory initiatives do not have a priori practical connections with real institutional policy processes.

Science shops (see, for instance, Mulder et al. 2006; http://www.science-shops.org/) are another way of empowering citizens, providing them with independent scientific and technological advice as required for local issues, in particular. A science shop is a "unit that provides independent, participatory research support in response to concerns experienced by civil society" (Mulder 2006, 279). Most science shops are linked to universities and use the work of students under appropriate supervision to respond to civil society's (mostly nongovernmental organizations) needs.

Interactive webpages of leading scientific organizations offer opportunities to the general public to get involved in discussions about future directions of science; see, for instance, CNRS (http://www.cnrs.fr/), UK Royal Society (http://www.royalsoc.ac.uk/), etc.

These initiatives may seem to be at the forefront, but what remains unclear, however, is to what extent there is a culture of government acting on the findings of such activities.

Doing it with ICT

In the last decades, we have been addressing throughout our research projects the use of information and communication technologies (ICT) both to promote spaces for interaction among the publics concerned and the communication of science for public debate of policy-relevant scientific developments (see, for instance, De Marchi et al. 1998; Guimarães Pereira and O'Connor 1999; Guimarães Pereira et al. 2001, "ICT Tools" 2003, "TIDDD" 2003, 2005, 2006; O'Connor 1998, 2006; Rosa et al. 2008; De Sousa et al. 2008).

I illustrate the usage of ICT for public debate of policy-relevant science with a project called GOUVERNe[7], which aimed at the development and pilot implementation of a user-based, scientifically validated process and informatics product for the improved governance of groundwater resources. In this project, our activities consisted of organizing and mediating the available knowledge about two groundwater resources case studies in Europe (see Guimarães Pereira et al., "ICT Tools" 2003; "TIDDD"2003).

The methodology deployed was based on the concept of *quality assurance by extended peer review* as a normative procedure to construct the knowledge base upon which a debate about water governance options could start in both case studies among the relevant social actors. What we called the "GOUVERNe process" was strongly based on transdisciplinary principles, combining hybrid methodologies and integrating social research methods with evaluation tools, such as multicriteria evaluation.

The GOUVERNe Process

Knowledge scrutiny in the GOUVERNe process was strongly based on social research. That was the means to ensure that knowledge other than scientific–technical was available in the knowledge base to debate on possible futures for groundwater resources and the associated river basins of the two case studies (in France and Greece).

The involvement of relevant social actors was done from the very first framing step, which ensured that, early in the process, their perspective, concerns and ways of representing the issues were accounted for. The research framing acknowledged and shared by the relevant community helps to avoid the so-called Type III error, of addressing the wrong problem, and enhances the scoping phase (i.e., focuses the work of collecting relevant information). The extended involvement also means that the issues addressed are shared and are dealt at the appropriate depth.

Clearly, if the experts involved in the process are the only "digesters" of the available knowledge (even if the process is *inclusionary*), their research framing and representation will be paramount. This is why the quality check by the relevant community throughout the whole process is essential for compliance and effectiveness, and why the process of creating socially robust knowledge (Gibbons 1999) is a continuous *inclusionary* process of those concerned. In GOUVERNe, engagement of relevant social actors was done at several steps of the process.

What emerged from the processes of knowledge scrutiny is that activities and options explored together by those concerned had great advantages in terms of enhancing the final process of dialog compared with those activities structured solely by "experts": what becomes available as a knowledge base to support the ongoing dialogs is perceived as *co-produced facts,* issues becoming more easily appropriated by all those engaged.

One of the main research issues of this process was how to *articulate* different values and perspectives, as well as different representations of knowledge that may be presented through alternative narratives (language, framing, scales of measurement, numerical models, etc). GOUVERNe is about knowledge integration while trying to keep diversity, which, in the interpretation of the researchers, was the creation of a space: where different types of knowledge articulated in different sets of semantic rules, with different codes, different scales of evaluation, etc., could be represented through several formats implying various degrees of specialization; where no a priori "integrative methodology" was applied as the means of sharing knowledge, the integration being made through dialog and interactions.

This entails the effort to produce a sort of "knowledge platform" that is accessible to all those involved and promotes conviviality of different *knowledges*, including tools that help with the process evaluation, capturing plurality and diversity and avoiding the pitfall of reducing them to something plausible but meaningless. This was explored through the use of ICT and, in particular, multimedia knowledge representation.

Building Spaces for Conviviality: The TIDDD Concept

A major development within the GOUVERNe process was the realization, design and prototype implementation of a new concept tool: TIDDD (Tools to Inform Debates, Dialogs and Deliberations) deploying new ICT. The main characteristics of this tool can be defined as "tools that inform and mediate processes of debate, dialog or deliberation which involve social actors of a governance, policy or decision process." Mediation of knowledge in this case entails organization, communication and exchange of a plurality of sources and types of knowledge (Guimarães Pereira et al., "TIDDD" 2003). In the case of GOUVERNe, there was a great deal of disciplinary knowledge, such as climate, geological and hydrological, as well as socioeconomical, regulatory, etc. Scenario drivers to debate about future options were devised together with the social actors. Hence, as some modeling tools were used to characterize possible futures, there had to be some work on "translating" that information in order to use it as input for the models. TIDDD's aim is the creation of convivial contexts of exploration and "discovery," where representations of knowledge come from different actors in the form of consistent narratives, aided by a multiplicity of supporting materials, namely multimedia formats, metaphors, etc. In TIDDD, some pieces of information were represented through different media in order to reach different people involved. TIDDD can integrate other sources and types of knowledge that may emerge during the process, which is done through the available multicriteria evaluation tool.

Quality assurance through extended peer review of TIDDD contents and design is one of the basic principles of this tool, since its main aim is to provide *socially robust knowledge* in contexts of societal debates and even scientific controversy. This is achieved through upstream engagement of the relevant community in the implementation of the knowledge base available in TIDDD, where the social actors check all developments and ensure that contents and design are suitable to start the debate on groundwater resources futures.

ICT for Public Debate

TIDDD-like tools were conceived as interfaces of mediation between policy spheres and other sectors of the society. This mediation is done with the help of *specialists*, but what comes out of the GOUVERNe process is that a new class of expertise is emerging, *specializing* in creating contexts for co-production of knowledge, in mediation of different types of knowledge, perspectives and values, and eventually *specialists* in making scattered, non-organized (for the minds of the experts!) pieces of relevant knowledge intelligible to the organized and oftentimes poorly flexible institutions: in a sense *transdisciplinary specialists* (Guimarães Pereira and Funtowicz 2005).

Involvement of the public in science and technology is about conviviality of different *knowledges*. It is hoped that TIDDD-like tools can help creating the "safe spaces" where "co-produced facts" can emerge.

"We have to Learn to Think in a New Way"

In the last sections of this chapter, we have first given a short historical perspective of how the public has been engaging with science, while also alluding to the fact that historical and contextual aspects of these relations still influence the ways in which the public relates to science today; science is, after all, a co-produced endeavor. Looking at more recent phases of how public involvement in science is conceived and fostered, and the still-prevalent "public understanding of science" paradigm, one can see that this recognition is just an implicit one. The recent survey of researchers and engineers done in the United Kingdom depicts actual ways in which many of them see their involvement with the public; i.e., such relationships are still embroidered within the latter mentality, concepts and practice. Hence, looking at frameworks and models of science and policy and science and society, explanations for the current state of affairs can be found, and, unsurprisingly, that PUS still shapes those relationships. Within such models and frameworks, calls for making explicit a relationship between science and its "interlocutors" based on cooperation and partnerships has been afoot for sometime. Whatever one calls this relationship ("extended," "co-production," "contextual," etc.) of the public having legitimacy to engage in the production of science, we are essentially saying that research has to be done in a different way. While discussing the *operationalization* of this engagement, we identify issues of political, institutional, organizational, cultural, civic, methodological, etc., that are at the basis of full embedment of public engagement in research practice, when such research is relevant for policy making. While exploring the usage of ICT, its is also noted that despite their potential both as connecting different *knowledges* and their "producers" and as a "safe" space for co-production of facts, we realize that ICT may challenge, but will not solve, the established research practices if changes in the researchers' mentalities and attitudes, and research culture in science–society relationships, do not occur.

And this leads to the title of this section: "We have to learn to think in a new way." In a recent workshop on "Moving Worldviews", Prof. Dürr from the Max Plank Institute challenged the ways the majority of universities still operate (Dürr 2005). Academics, especially those in technology studies, often leave the university thinking that they now possess the crafts and are now skilled to solve the problems of the world! They are now part of the "rationality crew" that will develop arguments and evidence to inspire and justify the ways in which the world operates. Often, this translates into a great deal of (perhaps unconscious) arrogance toward other *knowledges* and toward less-specialized fellow humans. This sort of "arrogance" is indoctrinated in staff and students at the universities as a means of protection, but it is also essentially a "way of being"—in Portuguese, we would say a *forma de estar*. Very early, the mission-to-be taught to "universitaires" is also a didactic one, of *training, educating* others.

Instead, more than ever, the approach is of humility—humility about both the limits of scientific knowledge and about when to stop *scientisizing* all *problematiques*. Jasanoff (2007) proposes that "technologies[8] of humility" are necessary to reflect on the sources of ambiguity, indeterminacy and complexity, and to accommodate dissent as Leach (2007) suggests. I am unsure about the use of the word "technologies" here. Rather, I view humility as primarily a guiding concept, which shall then guide a change of mentalities; in public engagement this is a shift towards *conviviality*.

This change would encompass the embedding of "technologies of engagement" as a natural step in research practice. The relationships of *specialists* with the public would go from a didactic mission to an engagement one. This will have to go hand in hand with learning how to establish dialogs, knowledge mediation, "integration of *knowledges*" and other "technologies" to make the "engagement" paradigm operational.

NOTES

1. The word *translation* is probably not the most appropriate since there is a great deal of "interpretation" in an activity of communication, which goes beyond the seeking of adequate vocabulary and tries to keep to the original semantics.
2. See, for instance, S. Miller, P. Caro, V. Koulaidis, W. Staveloz and R. Vargas. Report from the Expert Group Benchmarking the Promotion of RTD Culture and Public Understanding of Science (European Commission, 2002).
3. One of the most authoritative statements on PUS comes from the reports of the UK Royal Society, on *Public Understanding of Science*, published in 1985.
4. Bovine spongiform encephalopathy—commonly known as "mad cow disease."
5. The Royal Society and The Royal Academy of Engineering. Nanoscience and Nanotechnologies: Opportunities and Uncertainties (July 2004).
6. Wynne et al. (2007) alert to the extensive use of the stakeholder model as an ideal form of societal participation, which excludes a broader vision of citizenship. The term *stakeholder involvement* implies that the issues "at stake" are already agreed. Citizen involvement, on the other hand, carries no such presumption, and thus more readily accommodates diversities of "local" cultures with different preoccupations and concerns, meanings and worldviews.

7. Project no. EVK1–1999–00032: A Shared Cost Action financed by DG RTD, under FP 5. GOUVERNe stands for Guidelines for the Organisation, Use and Validation of information systems for Evaluating aquifer Resources and Needs. Online. Available at: http://neptune.c3ed.uvsq.fr/gouverne/ and http://kam.jrc.it/gouverne.

8. Technologies are inherently human artifacts, and here we would like to stress the importance of changing the nature of the human "posture"; in this case, the attitude of specialists toward nonspecialists and toward problem solving.

BIBLIOGRAPHY

AEBC. 2003. GM Nation? The findings of the Public Debate. Report of the AEBC. Available online at http://www.aebc.gov.uk/aebc/reports/gm_nation_report_final.pdf (last accessed July 14, 2008).

Bellamy, J.A., and A.P. Dale. 2000. *Evaluation of the Central Highlands Regional Resource Use Planning Project: A synthesis of findings.* Final Report to LWRRDC, Project CTC13, CSIRO Sustainable Ecosystems, Brisbane, November 2000. Available online at http://irum.sl.csiro.au/

Bellamy, J., B. Bishop, and A. Browne. 2004. *Ord Bonaparte Program Evaluation: Process and Impact.* OBP Project 1.3, Final Report, CSIRO Sustainable Ecosystems, Brisbane.

Burgess, J., and J. Clark. 2006. Evaluating public and stakeholder engagement strategies in environmental governance. In *Interfaces between science and society,* ed. Â. Guimarães Pereira, S. Guedes Vaz, and S. Tognetti, 225–52. Sheffield, UK: Greenleaf Publishers.

Callon, M., P. Lascoumes, and Y. Barthe. 2001. *Agir dans un monde incertain. Essai sur la démocratie technique.* Paris: Le Seuil.

CEC (Commission of the European Communities). 1997. Council Directive 97/11/EC of March 3, 1997 amending Directive 85/337/EEC on the assessment of the effects of certain public and private projects on the environment. *Official Journal OJ L 327,* March 14, 1997.

———. 2000. Directive 2000/60/EC of the European Parliament and of the Council establishing a framework for the community action in the field of water policy. *Official Journal (OJ L 327),* December 22, 2000.

———. 2001a. Democratising expertise and establishing scientific reference systems. Document of 2/07/2001. Available online at http://ec.europa.eu/governance/areas/group2/report_en.pdf (last accessed July 14, 2008).

———. 2001b. European Governance: A White Paper; Com (2001) 428, Brussels, July 25, 2001. Available online at http://eur-lex.europa.eu/LexUriServ/site/en/com/2001/com2001_0428en01.pdf (last accessed July 14, 2008).

Cortner, H.J., and M.A. Moote 1999. *The politics of ecosystem management.* Washington, DC: Island Press.

De Marchi, B., S. Funtowicz, C. Gough, Â. Guimarães Pereira, and E. Rota. 1998. *The ULYSSES Voyage: The ULYSSES Project at the JRC.* European Commission, EUR 17760 EN.

De Marchi, B., S. Funtowicz, and Â. Guimarães Pereira. 2001. From the right to be informed to the right to participate: Responding to the evolution of the European legislation with ICT. *International Journal of Environment and Pollution* 15(1): 1–21.

Duran, J., A. Hansen, and M. Bauer. 1996. Public understanding of the new genetics. In *The trouble helix—Social and psychological implications of the new human genetics,* ed. T. Marteau and M. Richards. Cambridge: Cambridge Univ. Press.

Dürr, H-P (2005) *We have to learn to think in a new way*. Paper presented at the *Moving Worldviews* workshop, The Netherlands, Soesterberg, 28-30 November 2005. Available on-line at: http://www.movingworlds.net/Downloads/Papers/ Dürr.pdf. Last accessed 14th May 2009.

Evans, G., and J. Durant. 1995. The relationship between knowledge and attitudes in the public understanding of science in Britain. *Public Understanding of Science* 4(1): 57–74.

Felt, U. 2002. Sciences, science studies and their publics: speculating on future relations. In *Social studies of science and technology: Looking back, ahead. Yearbook of the sociology of sciences*, ed. H. Nowotny and B. Joerges, 11–31. Dordrecht, The Netherlands: Reidel.

Frewer, C., G. Rowe, R. Marsh, and C. Reynolds. 2001. *Public Participation Methods: Evolving and Operationalising an Evaluation Framework*. Report prepared for the Department of Health and Health and Safety Executive, Institute of Food Research, Norwich, CT.

Funtowicz, S. 2001. Peer review and quality control. In *International encyclopaedia of the social and behavioural sciences*. Elsevier. 11,179–83.

Funtowicz, S. O. 2006. Why knowledge assessment? In *Interfaces between science and society*, ed. Â. Guimarães Pereira, , S. Guedes Vaz, and S. Tognetti, 138–45. Sheffield, UK: Greenleaf Publishers.

Funtowicz, S. O., and J. R. Ravetz. 1985. Three types of risk assessment: A methodological analysis. In *Risk analysis in the private sector*, ed. C. Wipple and V. Covello, 217–31. New York: Plenum.

———. 1990a. A new scientific methodology for global environmental issues. In *Ecological economics—The science and management of sustainability*, ed. R. Costanza, 137–52. New York: Columbia Univ. Press.

———. 1990b. *Uncertainty and quality in science for policy*. Dordrecht, The Netherlands: Kluwer Academic Press.

———. 1992. Three types of risk assessment and the emergence of post-normal science. In *Social theories of risk*, ed. S. Krimsky and D. Golding, 251–73.Westport, CT: Praeger.

———. 1993. Science for the post-normal age. *Futures* 25(7): 739–55.

Gaskel, G., N. Allum, M. Bauer, J. Jackson, S. Howard, and N. Lindsey. 2003. *Ambivalent GM Nation? Public Attitudes to Biotechnology in the UK, 1991– 2002*. Life Science in European Society Report: London School of Economics and Political Science.

Giampietro, M. 2003. *Multi-scale integrated analysis of agroecosystems*. CRC Press.

Gibbons, M. 1999. Science's new social contract with society. *Nature* 402: C81–4.

Glicken, J. 1999. Effective public involvement in public decisions. *Science Communication* 20(3): 298–327.

Gregory, J., and S. Miller. 2000. *Science in public, communication, culture and credibility*. Cambridge, MA: Basic Books.

Gross, A. G. 1994. The roles of rethoric in the public understanding of science. *Public Understanding of Science* 3: 3–23.

Guimarães Pereira, Â., S. Corral Quintana, S. Funtowicz, G. Gallopín, B. De Marchi, and B. Maltoni. 2001. *VISIONS—adventures into the future*. European Commission, Joint Research Centre, Ispra, Italy: EUR 19926 EN.

Guimarães Pereira, Â., and M. O'Connor. 1999. Information and communication technology and the popular appropriation of sustainability problems. *International Journal of Sustainable Development* 2(3): 411–24.

Guimarães Pereira, Â., J.D. Rinaudo, P. Jeffrey, J. Blasques, S. Corral Quintana, N. Courtois, S. Funtowicz, and V. Petit. 2003. ICT tools to support public participation in water resources governance & planning: Experiences from the design and testing of a multi-media platform. *Journal of Environmental Assessment Policy and Management* 5(3): 395–420.

Guimarães Pereira, Â., J. Blasques, S. Corral Quintana, and S. Funtowicz, S. 2003. *TIDDD—Tools To Inform Debates Dialogues & Deliberations. The GOU-VERNe Project at the JRC.* European Commission, Ispra, Italy: EUR 21880 EN.

Guimarães Pereira, Â., and T. de Sousa Pedrosa. 2005. V GAS©—Energy, Life-styles and Climate. European Commission, Joint Research Centre, Ispra, Italy: EUR 21869 EN.

Guimarães Pereira, Â., and S. Funtowicz. 2005. Quality assurance by extended peer review: Tools to inform debates, dialogues & deliberations. *Technikfolgen-abschätzung—Theorie und Praxis* 2(14):

Horlick-Jones, T., J. Walls, G. Rowe, N.F. Pidgeon, W. Poortinga, and T. O'Riordan. 2004. "A deliberative future? An independent evaluation of the GM Nation? Public debate about the possible commercialisation of transgenic crops in the UK, 2003." Understanding Risk working paper 04–02). Norwich: Centre for Environmental Risk. 1–182.

Involve. 2005. "People and Participation: How to Put Citizens at the Heart of Decision-Making." London: Involve and Together We Can. Available online at http://www.involve.org.uk/index.cfm?fuseaction=main.viewSection&intSectionID=400 (last accessed August 3, 2008):

IAP2. 2000. "Spectrum of Public Participation." Available online at http://www.iap2.org/associations/4748/files/spectrum.pdf (last accessed March 3, 2008).

Jasanoff, S. 2004. The idiom of co-production. In *States of knowledge: The co-production of science and social order*, ed. S. Jasanoff. New York: Routledge.

———. 2005. *Designs on nature: Science and democracy in Europe and the United States.* Princeton, NJ: Princeton Univ. Press.

———. 2007. Technologies of humility. *Nature* 450: 33.

Joly, J.-B., and A. Rip. 2007. A timely harvest. *Nature* 450: 174.

Miller, S. 2001. Public understanding of science at the crossroads. *Public Understanding of Science* 10(1): 115–20.

Morgan, D. L. 1998. *The focus group guidebook.* London: Sage Publications.

Mulder, Henk A. J., M.S. Jørgensen, L. Pricope, N. Steinhaus, and A. Valentin. 2006. Science shops as science–society interfaces. In *Interfaces between science and society*, ed. Â. Guimarães Pereira, S. Guedes Vaz, and S. Tognetti, 278–96. Sheffield, UK: Greenleaf Publishers.

Peals. 2003. *Teach yourself citizen juries. A handbook by the DIY jury steering group.* Univ. of Newcastle.

Pound, B., S. Snapp, C. McDougall, and A. Braun. 2003. *Managing Natural resources for sustainable livelihood. Uniting science and participation.* London: Earthscan Publications.

Rowe, G., T. Horlick-Jones, J. Walls, N. Pidgeon. 2005. Difficulties in evaluating public engagement: reflections on an evaluation of the UK GM Nation? Public debate about transgenic crops. *Public Understanding of Science* 14: 331–52.

Royal Society. 2006. "Survey of Factors Affecting Science Communication by Scientists and Engineers." Available online at http://www.royalsoc.ac.uk/downloaddoc.asp?id=3052

Slocum, N. 2003. *Participatory Methods Toolkit, A Practitioner's Manual.* Report published by King Baudouin Foundation & Flemish Institute for Science and Technology Assessment with United Nations University. Available online at http://www.kbs-frb.be or http://www.viwta.be

Taylor, P. L. 2007. Rules of engagement. *Nature* 450(8): 163–4.

Van Asselt, M.B.A., J. Mellors, N. Rijkens-Klomp, S.C.H. Greeuw, K.P.G. Molendijk, P.J. Beers, and P. van Notten. 2001. *Building Blocks for Participation in Integrated Assessment: A Review of Participatory Methods.* ICIS report I01-E003, Maastricht, The Netherlands.

Wilsdon, J., and R. Willis. 2004. *See-through science: Why public engagement needs to move upstream.* London: DEMOS.

Wondolleck, J.M. and S.L. Yaffee. 2000. *Making collaboration work: Lessons from innovation in natural resource management.* Washington, D.C.: Island Press

Wynne, B. 1995. Public understanding of science. In *Handbook of science and technology studies,* ed. S. Jasanoff, G.E. Markle, J.C. Petersen, and T. Pinch, 361–88. Thousand Oaks, CA: SAGE.

Wynne, B., U. Felt, M. Callon, M.E. Gonçalves, S. Jasanoff, M. Jepsen, Joly, P.-B., et al. 2007. Taking European Knowledge Society Seriously. European Commission: EUR 22700 EN.

3 Active Citizenship, a Case Study
The Controversy of High-Speed Rail in the Susa Valley

Luca Giunti and Elena Camino

EDITORS' INTRODUCTORY NOTES

This chapter presents a case study of a socioenvironmental controversy that is currently taking place in Italy. A proposal for a large-scale engineering construction—a high-speed railway connecting Italy and France [Treno ad Alta Velocità]—has been contested by local populations: over a period of time, a number of public protests have been organized, with some outbursts of open conflict. This case bears resemblances to many other controversies occurring in different countries, and it is illustrative of problematic conditions in public policy arising from the attempted or effective application of the most powerful means of modern technoscience for the transformation of natural systems. A detailed analysis of the controversy as it has developed over time is presented here to illustrate how the conflict between the parties involved is more than a quarrel generated by idiosyncratic and contrasting forms of evidence or by alternative interpretations of factual data, but rather it is a contradiction of aims and purposes, generated and sustained by the application of different ways of looking at reality, society and science.

In addition, the experiences of the Susa Valley feature aspects of creative conflict transformation: this aspect is presented here as a practical implementation of the links between the interpretive framework of post-normal science and the practice of nonviolence of Gandhian origin. After all, the idea of democratic participation flourished in the same historical and cultural context. The "TAV case" presented in this chapter has also been used in education as a stimulus for the development of a role-play, an educational strategy described in Chapter 8.

COMPLEX AND CONTROVERSIAL SOCIOENVIRONMENTAL ISSUES

Situations of conflict between institutional powers on the one hand and citizens' groups on the other are on the increase everywhere in the world.

According to Latour (2005), we now have a Realpolitik of laws and top-down decisions that is setting itself further apart from a Dingpolitik consisting of the voices and forms of organization of lay people, with different—and sometimes contrasting—interests and needs. Such a contrast can be found in those situations in which the quality and/or variety of natural services available to resident communities is endangered[1]. The construction of infrastructures for waste disposal, airports, traffic bypasses and large-scale constructions in general, necessitate radical alterations to the landscape, and exceptional movement and transformations of both matter and energy. In other words, these are high-powered interventions, and such activities are resisted by local populations not only because of the negative impact on their personal well-being, but more often because they are considered to be of little use or generally pernicious (Gallino 2005, 2006). Hence the not in my backyard (NIMBY) syndrome is being replaced in some circumstances by malaise and opposition toward a worldview that measures progress through economic growth.

THE EMERGENCE OF CONFLICT

This chapter reflects on this new political scenario by looking at the dynamics of a controversy that has gradually taken the form of a conflict between two groups, with each group consisting of people who take sides for a variety of different—not always completely shared—reasons. The controversy of high-speed rail in the Susa Valley in Northern Italy[2] illustrates a situation in which the relationships between opposing groups have been characterized by competitive attitudes: mutual mistrust, use of arguments in order to win and power dynamics, which have led to an impasse. In such conditions, groups of people divided into two adversarial fronts experience a hardening of their mutual positions, and by so doing, become less amenable to those processes of open and respectful dialog, or strong but constructive resistance, which could lead to a transformation of the conflict.

In illustrating this case, we find many resonances with previous chapters: Chapter 1 looked at technoscience and the impact of its application. Chapter 2 mentioned the people—citizens and local populations—who increasingly express their dissent toward decisions made by the political power and the experts, and the moves toward a more participatory democracy.

In particular, a third element, which is more explicitly mentioned here, is conflict: a conflict relationship (between individuals, groups, states) is usually interpreted by the Western culture in negative terms. In reality, conflict is one of the most common forms of relationship between humans and it can be experienced and managed not only through destructive means (with forms of direct and indirect violence), but also in a constructive way, and in this way proving to be a creative opportunity. Learning to reflect on conflict and acquire competences for analyzing it, managing it and, where possible, transforming it by nondestructive means is clearly of great importance in a scenario of sustainability.

NONVIOLENT CONFLICT TRANSFORMATION

Creative and nonviolent conflict transformation has been the subject of study and practice of Johan Galtung, one of the most prominent authors of Peace Studies, for many years. He has been involved in this field both at the theoretical and operational levels (Galtung 1996), having promoted and run many training courses, seminars and mediation meetings between groups and political delegations.

Galtung makes explicit reference to Gandhi when he proposes to replace the triangle of violence with the triangle of nonviolence. The triangle of violence is manifested in violent behaviors that are accompanied by feelings of hatred and resentment. In this case, the conflict is characterized by incompatible aims. In a triangle of nonviolence, the subjects are able to make a distinction between the people (for whom they may feel empathy) and the problems causing the conflict, drawing on their creativity to overcome the situation of impasse.

Explicit and extensive references to nonviolence and Gandhian thought are also found in Ravetz (2006, 281):

> Under what circumstances can there be a creative outcome of a characteristic contradiction in a social system? This requires the various actors and subsystems each to make some sacrifice for the common good. In a conflict situation (which is characteristic of such a state of affairs) each must recognise the humanity of the enemy whom they may, with good reason, believe to have done inhuman things. Hence the politics and philosophy of non-violence, best understood in the Gandhian term "satyagraha" (struggling for one's own truth) is no longer just a luxury sentiment of Western middle-class idealists. It may well lie at the heart of the transition to a sustainable civilisation.

The transformation of a real conflict can thus be seen as the practical and operational outlet of the need theorized by the proponents of post-normal science (Funtowicz and Ravetz 1993, 1994, 1999) and sustainability science (Gallopin et al. 2001) to multiply the number and variety of perspectives through which a complex system is being studied and to give all stakeholders' voices the opportunity to be expressed and heard. Using Galtung's words (Galtung 2000): "the more complex the conflict, the more numerous the openings for its nonviolent and creative transformation" (144).

Also, Giuliano Pontara (1973, 1996)—an attentive scholar of Gandhi's thought—stresses the importance during dialog of accepting the principle of fallibility. According to this principle, we are all mortal beings with limited powers of knowledge, hence nobody can ever be sure that what one believes at any moment to be true is in effect so—it may well be false. The principle of fallibility holds true in the first instance in the field of science, as we have seen in Chapter 1. Yet it is equally valid in the field of ethical belief:

> An individual who is provided with a nonviolent personality (. . .) will not want to exclude a priori the possibility of being wrong and the adversarial being right. For such reason, he/she rejects means for dealing with conflict that lead to the destruction of the other (. . .). Hence the internalisation of the principle of fallibility is one of the best immunisations against all forms of fanatic behaviour (. . .); it is also fundamental for the good functioning of democratic institutions and it constitutes a great incentive for tolerance (. . .) (Pontara 1996, 59–60).

The approaches of post-normal science and nonviolence thus provide the epistemological and philosophical basis for those forms of public participation described in Chapter 2: these are situations and contexts in which participants hold equal dignity, and communication between people and/or groups is founded on respectful dialog.

From this, two important implications follow. The "explanation" of a complex issue is intended as a dynamic process of interpretation and it is always transitory. This derives from a process of integration, elaboration and transformation of partial and somewhat contradictory perspectives. In this scenario, previously hardened positions can become more fluid and the adversarial set ups can be modified. In addition, the geographical and historical boundaries of the conflict can be opened, promoting the inclusion of new parties and the creative transformation of the scenario.

A CASE STUDY

The previous chapters provided the theoretical basis for understanding the complex transformations that occur in socioecological systems as a result of the production and application of technoscientific power. Both chapters make evident the close interconnections between science, technology and society in a world that is increasingly more interconnected, and, for such reasons, more exposed to conflict.

This chapter sets out to illustrate such complex interactions in a case study. In particular, the issue of a high-speed rail link connecting Italy and France through the Susa Valley represents a conflict centered on the management of natural resources and described by a strong component of ecological equity. In addition, it introduces themes that will be dealt with in the following chapters and related to ways of knowing (Chapter 4), possibilities for change and the role of formative processes for a "sustainable" education (Part 2).

The Issue of High-speed Rail as an Emblematic Case

The Susa Valley—A "Corridor" Filled with Variety

The Susa Valley is located in Northwest Italy at the border with France, from which it is separated by the Alps, 3600 meters high. It is the widest valley in the Western Alps; in fact, it is a natural corridor stretching from East to West. The two sides of the valley benefit from different sun exposure and this makes them quite different from one another. The left side is dry, while the right side is humid, shady and cold. The natural environment, and particularly the flora, are deeply affected by this peculiarity, resulting in a valley with extremely variegated and interesting sites and habitats.

The Dora Riparia River runs through the valley, and there are abundant springs and superficial aquifers. In the high part of the valley there are pastures, while at lower heights (1300–1800 meters) there are steep crevasses.

The Susa Valley is the most developed of the alpine valleys. It hosts two main roads directed toward two international passes (Monginevro and Moncenisio); a motorway and an international railway, each one with a tunnel at the Fréjus; two electric lines and three hydroelectric dams; many tourist and sport resorts (the valley was the base of the 2006 Winter Olympics). There are many industries, including mining, and many military roads built in previous centuries that are currently international tourist attractions for walkers and cyclists.

The valley has about 90,000 inhabitants, and it is divided into 39 councils. There is a well-established tourist industry, as is evident by the "second" homes, hotels and heavy motorway traffic. Notwithstanding the heavy human presence, the Susa Valley features wide seminatural and wild areas, which host many examples of alpine fauna (deer, chamois, roe deer, wild boar, eagles, hawk, partridges and wolves) and a very rich diversity of flower species: there are four natural parks, two natural reserves and many sites of communitarian interest[3].

Sheep rearing, which was flourishing until the end of World War II and subsequently in decline, is now in a new phase of growth, albeit slow, and it consists of about 7000 cattle, 10,000 sheep and 500 goats.

The Susa Valley has a very ancient history, and many signs of the past are still visible: archeological sites, Roman villas, churches and abbeys, and castles and fortresses, which attract thousands of tourists every year.

At the time of the Romans, the alpine crossovers of the valley acquired strategic and military importance: it is believed that Hannibal trespassed in

218 BC and Julius Caesar crossed over in 61 and 58 BC, marching towards Gallia. After the fall of the Roman Empire, different populations succeeded each other: Goths, Byzantines and Longobards, and in 774 ACE, Charles Magnum descended to Italy from the Susa Valley.

Toward the end of the twelfth century, abbeys were founded that are well preserved and open to visitors today. Between 1600 and 1800, the Susa Valley was crossed by many armies. In 1854, the Turin–Susa Railway was opened, followed in 1871 by the Fréjus tunnel. After the World War II, the valley remained part of Italy, but the Valle Stretta and the Moncenisio were handed over to France.

In conclusion, the geographical location of the Susa Valley makes it a site of extreme wealth of natural resources, which have supported economic and cultural development over many centuries. However, the history of the valley also suggests that this has always been an area of conquests, conflicts and political appetites.

The High-speed Rail Project and the Evolution of the Conflict

The Turin–Lyon high-speed rail line is part of the "Trans European Network," which has been planned by the European Union for the next 20–30 years. It is a section of the so-called 6th Axis, which should, by means of one single rail track, connect Portugal with Ukraine from Lisbon to Kiev and across central Europe. A rail connection already exists, but the technical characteristics of the railways in each country are different and the trains' speed is thus reduced. For such reasons, a brand new line, modeled on the French high-speed trains, was planned. This would consist of dedicated tracks and infrastructures, separated from previously existing lines. A fundamental point in the project is concerned with the passage through the Alps, which—according to the proposers—should involve the Susa Valley and the construction of a 52-km-long tunnel.

The First Signs of Conflict

The controversy originated at the beginning of the nineties with the formation of the committee proposing the construction of the Turin–Lyon high-speed rail link. The committee was originally constituted by a group of industrial people. A year later, the Committee Habitat, consisting of about 60 members including technicians, doctors, professionals, factory workers, lecturers of the Turin Politechnic, mayors and valley civil servants, was formed. In 1994, after studies and discussions that lasted a few years, the Committee Habitat expressed a negative view of the high-speed rail, marking the start of the conflict. In the following years, long pauses, during which it seemed that the conflict had settled, alternated with episodes of resumption: a few Molotov bombs in the Susa Valley; activation of a national network against high-speed trains; meetings and conferences on "High Speed and shady businesses."

Between 1999 and 2004, the summits amongst the proponents of the high-speed trains multiplied. In parallel, meetings and demonstrations against the high-speed trains also grew in number. The local administrators of the Susa Valley took sides with the "NOTAV" movement and had to deal with the local and national representatives who had been taking inconsistent positions: in 1999, the ministry of the environment declared, "Forget the High Speed rail, it won't happen"; a short time after, the President of the Piedmont region and the mayor of Turin earmarked the high-speed Turin–Lyon train line as a priority. The conflict gradually grew in intensity: gatherings, public demonstrations and roadblocks accompanied the spreading of documentary information related to the network and the works that had been foreseen. The European Commission put the Turin–Lyon high-speed rail among the priority items to be funded and this decision triggered a public demonstration organized by 38 mayors and the President of the Comunità Montana. This was followed by another rally, which included 37,000 people.

The opposing forces, Pro and Against the construction of the new railway track, consisted of the Government and regional management organs on the one side, and citizens' organizations and nongovernmental organizations on the other side. The different voices are listed in Table 3.1.

Table 3.1 Voices in Favor of and Against the Construction of the High-Speed Rail Track in the Susa Valley

• CIG (Commissione Inter-Governativa; president: Sergio Pininfarina)
• Italian Government (both right-centered and left-centered)
• Railways (Italian FS and French SNCF)
• Comitato TRANSPADANA: three banks, six entrepreneurial federations, five Chambers of Commerce, two Provinces (Turin and Trieste) , six Communes (Genova, Torino, Milano, Brescia, Verona, Trieste)
• Piedmont Region, Turin City and Shire
• Turin Industrial Union with companies such as FIAT
• CRT Bank, St. Paul Bank
• Associations: Turin International, Rotary Club, Lyons Club
• Articles published by Mass Media: La Stampa, La Repubblica, TG3 Piemonte

• People's committees of resistance "NO-TAV":
– of Susa Valley (Bussoleno, Condove, Caprie, Almese, Avigliana . . .)
– of "Gronda" (Val della Torre, Pianezza, Savonera, Venaria, Turin . . .)
• Associations: Habitat, Pro Natura, Legambiente, Valsusa Filmfest . . .
• Institutional Committee for high-speed rail, including:
– Comunità Montana Bassa Val Susa and single authorities
– NW Gronda Communes of Turin
– Coldiretti Social Forum and Networks and Associations, Committee NOTAV Turin
• Local parties: PRC and Verdi
• Articles published by local media: Luna Nuova, La Valsusa, Dialogo in Valle

The Escalation of the Conflict

Between 2004 and 2006, there were direct clashes between the parties: technicians from the Italian Rail Network attempted to open a building site in order to undertake geognostic measures, but they were obstructed by the local people. Subsequently, after having sent letters of reposession to some inhabitants of Venaus, a small village in the high Susa Valley, technicians from Lyon–Turin Ferroviaire (LTF), escorted by the police, tried to reach three testing sites. They were forced to go back by the NOTAV people. One of the sites was fenced off overnight; this act triggered further protest in the days that followed. On the night of November 29, 2005, hundreds of men and military forces reached Venaus and surrounded the area, impeding access to anyone. Local people and police confronted each other for the entire day at the Venaus crossroad. In the evening, the block was dispersed and the people reached the area of the future building site, which was defended without interruption.

On the night of December 6th, the police occupied the area of Venaus, violently forcing back the people and using caterpillars to take down tents, barricades and shelters. The technicians fenced off the land with nets. In the following days, 50,000 people marched from Susa toward Venaus, headed by the town mayors. Once they went past the roadblocks, they reached Venaus and took down the fences, regaining their land. The government called the mayors to Rome and withdrew most of the troops. Many more actions of protest followed in December: demonstrations, concerts and congresses.

The agreement of mutual help, the local attempts at dialog and the initiatives from the top.

The demonstrations continued, not only at the local level, but also in connection with other similar situations of protest against big construction enterprises planned in other regions in Italy. A "Patto di Mutuo Soccorso" brought together all those who were opposed to big construction projects: the NOTAV movement took part in rallies in Vicenza to support a protest against a US base, and in Sicily, against the oil rigs. They were present in Piedmont to demonstrate against the nuclear waste, and in Naples to participate in a national rally on the waste emergency.

In September 2006, the work of the Turin–Lyon *Osservatorio* began. This was an official discussion forum opened to all stakeholders with the purpose of considering the shared understandings and getting the controversy under control. It is anticipated that the documents produced during the discussion will be published. In the meantime, the current Italian government has repeated its intentions to build the Turin–Lyon railway.

While the NOTAV movement has gathered 40,000 signatures with the petition "Against TAV without if's and without but's" and handed them over to the European Parliament, the Italian government has applied for a grant of 750 million Euros from the European Parliament for the construction of the Turin–Lyon rail link. A preliminary project has been presented with a new track for the tunnel.

And Now?

Almost 20 years have passed since the start of the conflict. The documentation that has been produced about the project is extensive and it is clear to all those involved that this is an extremely complex issue for which—according to Funtowicz and Ravetz (1999)—the uncertainty of the system is evident and the stakes are high. An immediate example of this condition can be taken from the consequences of the recent increase in oil prices: in May 2008, the European Union had estimated that the projected expenditures for the railway line Lyon–Turin–Trieste–Lubjana had already been subjected to an increase of 38.2%. Equally unpredictable at this time are the political events in Italy. The TAV project (as it often happens for all large-scale constructions) is characterized by progressive concentration of private financial, economic and political power, and as it was indicated in Chapter 1, uncertainty and ignorance are not accidental and temporary conditions, but they are intrinsically and radically embedded in the matters to face.

The NOTAV movement is fatigued after 20 years of resistance, and internal tensions are mounting. Many people have lost trust in the efficacy of the Osservatorio. Some people would like to reach a conclusion, even a compromise; other people are concerned about the economic situation, and they would be willing to accept some kind of compensation and let the work start.

One part of the movement, however, has produced documentation, written books and information material, and has undertaken creative actions: the most recent one is the campaign "Buy a seat in the front row." Whoever signs up to this initiative buys, at a symbolic price, a square meter of the land through which it is planned to build the high-speed rail. The aim is that of obliging the proponents to carry out hundreds of expropriation practices and to be present in legal and democratic forms at the start of the construction works. The initiative has gathered more than 2800 signatures (August 2008).

The Elements of the Controversy

The main points of divergence of opinion about the issue of the high-speed rail have been widely debated in pubic meetings, documents and books: for example, Bobbio 2006; Campana et al. 2006; Della Porta and Piazza 2008, among many. The problematic aspects can be summarized in four typologies: economic, political, methodological and environmental. Let's look at them in detail.

Economic Aspects

The anticipated costs for the construction are large, and previous experiences in Europe and overseas resulted in costs up to 40% above the original estimate. In addition, the other high-speed rail tracks in Italy cost an average of 30% more than Spain or France. The expenses will be largely covered by public money, national or European. Hence, a very accurate and in-depth assessment of the economic sustainability of the project is to be demanded, along with a cost–benefit analysis to be documented and shared among all stakeholders. Some critics of the TAV are from the economists' front: "it is necessary . . . to finally take the route of the correct economic assessment of projects, in order to select the real priorities" (Boitani and Ponti 2006, 108).

The project's supporters stress the urgency of the situation: every delay is a potential loss of opportunities for financial support from the European Union. One of the benefits that is envisaged from the project are job opportunities at the building sites; those who are against the project, however, fear the possibility of unlawful manipulations in the calls for tender, and remember that this is what had already happened for the Winter Olympic games in 2006, when most of the workforce had been called in from other countries.

There are also significant divergences of opinion about the anticipated rate of use of the high-speed train with respect to transport of goods and passengers: some people maintain that freight transport is in decline, while others hypothesize that it is on the increase. Some economists fear that, if the line is not built, Italy will be "cut off" from Europe; others stress that Italy is sufficiently integrated and that there is no need for additional structures.

Political Aspects

The decision to build a large-scale construction such as the Turin–Lyon high-speed rail link (as with the bridge on the Messina canal, the tunnel under the Channel, the Oresund Bridge between Denmark and Sweden and so forth is certainly a political decision because of the anticipated future, the high investments, the territorial impact, the environmental and occupational effects and the involvement of a plurality of different subjects (public administration, companies and enterprises from various sectors). It is at this level—the political one—where different worldviews clash. Often in macroconflicts "what is most important are the deep structures and cultures, because these are not reflected upon, they are even unconscious" (Galtung 2000, 481).

In political discussion there are people who argue for the usefulness of a new train line and people who maintain that it will be useless. The NOTAV movement accuses politicians of favoring specific economic interests or of giving into the pressures of entrepreneurial power. The supporters underline the need to reduce road traffic in order to respond to the needs of the local population for a healthier environment. Those who accuse the government of not having clear policies on transport are presented with a scheme for a European rail network, of which the TAV is an integral part.

Environmental Aspects

That the TAV project will have a great impact on the territory is acknowledged by everybody. The environmental concerns are mainly related to the possibility of coming across asbestos- and uranium-rich rocks during the excavation works, the possible interference with water aquifers, air and water pollution and the occupation of green land, which is partly cultivated and partly wild. The supporters of the project accuse the opposition of being affected by the NIMBY syndrome; they stress the advantages of a reduction in road transport and they are adamant that modern technology will be able to solve the possible damaging effects derived from the excavation of asbestos-rich rocks. Thanks to the presence of expert technicians, they are sure of being able to redirect the water, which might be encountered during the excavation works, to the inhabitants of the valley (and more comfortably than would previously have been possible). The TAV promoters do not deny that the inhabitants of the Susa Valley will have to undergo some discomfort for a few years; however, they reckon that the economic compensations that have been foreseen will provide sufficient recompense. They will be able to manage protected areas more effectively and put infrastructures in place to increase the tourist intake. Recently, the president of the region stated (2007): "the Pianura Padana[4] is a continuous, endless line of trucks. We need to choose a different model of development and to engage in the completion of the high speed rail line in the Pianura Padana".[5]

The opposition is sceptical: environmental disasters that have recently occurred in similar situations make them worry about the possibility of damaging the underground water system. The prospective of seeing the valley turned in a big building site for about 20 years is worrisome because of the environmental impact that they foresee will be felt both by the human populations and the natural environments, which have so far been attentively safeguarded.

Methodological Aspects

The proposal to build the high-speed Turin–Lyon rail link has been presented more than once at government level and confirmed—by both the left and right wing—without any preliminary involvement of the local administrations and without listening to the local voices. During the 20th century until the beginning of the 1990s, infrastructure projects of national interests have been put in place using a top-down model: the political decision was accompanied by expert technical planning, which had been requested by the government. With the increasing scale of the projects—both in financial and environmental terms—this way of working has encountered mounting opposition by the people and sometimes also by the local administrators, creating not only conflict between institutions and citizens, but also between central and local institutions. It is not only about reclaiming the abstract right of being involved: in this case, the protest also involves the validity of the data that have been found to be largely incomplete and, in many respects, unreliable. Many members of the NOTAV movement—not only those who could be recognized as the "experts" because of their profession (for example lecturers at the polytechnic), but also many inhabitants from the affected areas having direct experience of the territory—have contested the project since the beginning and have offered their competences in order to provide a broader framing of the problem. In this respect, we can refer back to Chapter 1, where it was stated that:

> in the context of complex and controversial socio-environmental issues (. . .) new forms of public control of knowledge production have to be elaborated (. . .) this entails the necessity, primarily epistemic and methodological, to extend the public participation in decision making processes (Chapter 1, 20).

In practice, this is translated into the realization of a "contextual model": "Public are not passive recipients of knowledge inputs provided by specialists, but are able to engage in 'co-production' of facts" (Chapter 2, 38). Moreover, in this case, the inhabitants of the Susa Valley can be considered citizens who "(. . .) derive much relevant knowledge from everyday experience, dealing with real world problems" (Chapter 2, 40).

From their side, the proponents of the project have always maintained to have taken care of the communication with the public, by means of providing reader-friendly documentation, often through the institutions[6], dedicating ample spaces for discussion in newspapers and television programs. They have tried to use a common language that could be understood by a nonexpert pubic. This position can be related to an idea of science as "provided with a privileged epistemological status" (Chapter 1), and to the approach of the Public Understanding of Science, illustrated in Chapter 2.

In addition, with regard to the issue of the TAV, the alternating of different governmental forces has made the issue even more intractable. Although each constituency has always supported the project, the ways in which the political processes had been developed have been quite diverse. More specifically, while with one governmental constituency, a law of environmental impact assessment (EIA) had been requested, with the subsequent constituency, a "target law" has been advanced. According to the declarations of the President of the cabinet, this law will turn the country into a large building site: "by means of the target law which will accelerate the planning phase of large infrastructures and will open a preferential lane to strategic projects, we will be able to reduce the delay with the other countries." The application of the target law will involve a reform of the procedures for EIA and environmental authorization. This more radical position taken by the government has contributed toward hardening the position of the opposition movement. This has also been enriched by the participation of local administrators, who found themselves in the difficult role of representatives of the institutions but holders of opposing positions.

The Interviews with the Protagonists

As part of a research project in science education conducted by a few members of our research group, some of the people who have been directly involved in the controversy have been interviewed: technicians, administrators, local inhabitants of the valley and civil servants. The research was aimed at casting light on the perceptions of the people involved and their level of competence in describing the conflict (the various levels at which the conflict had occurred, the people or groups with a stake in the conflict, the reasons for contention), as well as their ability to define the boundaries of the context, both in time and space and to imagine possible future scenarios. As explained in the letter that was sent to the people who were asked to be interviewed:

> the research is aimed at analysing the linkages between situations of conflict connected to environmental issues (access to resources, land use, ecological footprint) and more general intercultural conflicts (between different ethnic groups, between genders, developed industrial countries and depressed areas etc . . .). [. . .] Specific attention will be paid to situations in which nonviolent means for dealing with environmental and intercultural conflict have been (or could be) adopted (Camino et al., 2008a).

The interviews (18 in total) and the subsequent data analysis showed the different degrees of familiarity of the interviewees with the more reflective and implicit aspects of the issue: the theory and practice of dealing with conflict, the different epistemologies of science, the variety of possible interpretations of science–society relationships. We came across citizens who—each one in his or her respective domain of competence—argued their own position on the TAV issue with a wealth of data and linguistic competence. From the answers that were gathered to the question, *"What type of conflict do you think is characterising the 'TAV/TAC case'?"*, it emerged that people lack a common vision of the reasons for contention. Here, we report some of the answers:

- Conflict between competences, models of development and local people, between local and central administration (Piedmont region civil servant).
- It is a typical example of NIMBY. The debate arises when the construction involves a collective group; an intervention is planned for the benefit of the collective, but it is the locals who bear the cost (technician of a public administration).

- The conflict was generated by the differences and clashes between the official data on environmental assessment and the evidence provided by the consultants of the Comunità Montana (Administrator of the Susa Valley).
- It is a shortcoming of the governance, of the group of people who have to make decisions (inhabitant of the valley).

Moreover, we ascertained that there is wide divergence with regard to the most appropriate decisions to be taken in the case of acute conflict. Here are some of the answers to the question, *"In case of acute conflict, what kind of decision-making procedures do you think are most appropriate?"*:

- If the decision is that of building the line, then this is to be done, with the army of thereabout.
- To stop. To avoid direct clashes and to reinstate dialog. Then create a space for discussion without starting from the assumption that one party has to convince the other.
- Perhaps it is the task of an independent authority to make a decision after having listened to all stakeholders.

Also, in relation to the geographic boundaries of the areas that are potentially affected by the construction works, different ideas were expressed: a few people reckon that the impacts are limited to the directly affected region; the majority does not hesitate in defining a global context: "we are no longer in the position of thinking about local scenarios" (one inhabitant of the Susa Valley).

On average, the interviewees displayed a limited ability to grasp and define the complexity of the socioecological system and some difficulty in imagining (by describing and comparing) future scenarios, both within the boundaries of the local context and the international arena. A widespread feeling of unease was expressed about the ways in which the government handled the controversy, although this was not always accompanied by concrete suggestions for alternative strategies. From the picture that emerged, it appeared that a shared "method" for tackling the problem was lacking. There was no evidence of specific competences for grasping the complexity of socioenvironmental systems in a rigorous and systematic way. For example, by making reference to the joint presence of multiple time and space scales; nonlinear interactions and feedbacks between different organizational levels; plurality of points of view and so forth. Differences of views are expressed about the most appropriate ways for organizing roundtables and initiating the decision-making process, although without acknowledging such a disparity of views as a problematic aspect. For example, some people supported the idea of delegating responsibility on matters of public decisions, whereas others reckoned that forms of participatory democracy are essential. This is an aspect that can multiply the reasons for contention by creating confusion between issues of method and issues of merit.

The interviews mainly express firmly opposed positions; the controversy is perceived by the majority through a dualistic framework (YES/NO) that overlooks the complexity of the problem and the multiplicity of levels and perspectives, which should be appropriately considered in order to understand it. Only a small minority express competences for a nonviolent transformation of the conflict. The scarce familiarity with the culture of nonviolence is also displayed through the difficulty encountered by many people in becoming aware and expressing their needs, interests and worldviews more clearly, as this could be a means for making more explicit the real reasons for contention (Camino and Dogliotti 2004).

In this regard, among the many comments that have been made in the past by scholars, writers, experts etc., it is worth mentioning a reflection expressed by Luciano Gallino, a sociologist who is well recognized in Italy:

> (. . .) perhaps there might be two routes to be taken for a calm appraisal of the TAV issue. One is to recognise an unsolvable, fundamental conflict between the local interests and the national interest [. . .]. The second one however is so much wider, considering the economic advantages that the project promises to bring to the nation, that a decision in favour of its construction is almost inevitable [. . .]. Alternatively, a second route could be that of adopting a different mindset which

"He says he really does not know where to put them, all the trucks filled with spoil"

Figure 3.1 What goes around comes around: Trucks filled with spoil.

begins with asking ourselves if perhaps it is the inhabitants of the Susa Valley who by means of their opposition to the TAV project are serving the national interest [. . .]. It is not only a rhetorical exercise to put two different mindsets into dialog to see where they could lead to. (Gallino 2006, 42–3).

The Nonviolent Component of the NOTAV Movement

The NOTAV movement has, since the beginning, adopted a nonviolent position (Cheli 2002; Patfoort 2006; Salio 2006), helped by the fact that, since the early seventies, a group of Gandhian origin had been present and active in the valley. The occupation of the sites that are the object of controversy is in line with a nonviolent approach: such sites become places for gatherings, meetings, debates, conferences and concerts. All the layers of the population are involved: the initiatives are attended by teachers and students, factory workers and white collars, pensioners and housewives, doctors and firemen, priests, families. The nature of the movement that is nonviolent, assertive but open to dialog, is well documented by witnesses (such as one of the authors of this chapter, Luca Giunti) who have been present on the garrison houses many times, interacting both with the local inhabitants and the police. We give here some examples of what has been happening:

> During the clashes at the Seghino, the carabinieri[7] grab a boy who is picturesque and pierced but fairly innocuous and good natured. They give him a telling off and then put him in the van. They are not familiar with the place though. Hence the boy, gently, points them to the road for the Susa police station and he takes them there!
>
> Two ladies, well into their fifties, are having a look around. They take notice of some garbage lying around. They grab two black bags and start picking up the waste so that "the battlefield is clean". A policeman over the fence sees them and asks for a black bag because "also here behind us there is a lot of garbage". The rubbish bag is promptly handed over.
>
> The fourth consecutive day I reach the garrison in the morning. I park the car and I walk past the car of the carabinieri which is blocking the road. It is very cold as the sun never reaches this point. I wave to the carabinieri inside the car. They respond with a bright "buongiorno!" which I return. It feels like being in the cartoon of Wile E. Coyote and the sheep dog, which clock in together with a warm hello and a minute later they fight . . .

Many episodes that occurred during the conflict—which is still ongoing between the promoters of the high-speed rail and its opponents—have been interpreted within a perspective of nonviolent resistance: "We can see the breaking in of the police as one of the many examples of State violence, an institutional one which is practiced in the course of nonviolent actions (from Gandhi to Martin Luther King and Nelson Mandela) and which can trigger the phenomenon of the political ju-jitsu, with a boomerang effect towards those who make use of such violence. The first effect of the police attacks was the extension of solidarity to a large number of citizens' committees and base groups, associations and some political forces, from Turin to Milan, from the Alps to Sicily (Salio[8]). Nanni Salio continues by saying:

> We also know that your opposition is not only a reaction 'against' and not even simply concerned with personal, local interests. You have proposed and developed good alternatives, what Gandhi would have called "constructive" programs.

In fact, the NOTAV movement has so far showed good doses of creativity, sense of humor and determination, for example, by promoting the publication of books and reviews (for example, Bettini and Cancelli 2006; Mercalli and Sasso 2004), popular materials (such as the comic "Asterix and the battle of Venaus" 2005), and most of all by inventing effective actions that could involve a growing number of people while at the same time developing "resistance" toward the adversaries. The most recent initiative—mentioned earlier—is that of promoting the buying of land plots in the area of conflict. Here is how the news has been reported by newspapers and magazines:

La Stampa, Turin, February 10, 2008:

> The Committees have decided to buy "a section of the land which will be the site of the construction works". Thus "thousand, five thousand, ten thousand land proprietors will be chosen like leopards spots where extraditions and temporary occupations will have to take place". The consequences? The company which will have to put the building site in place will be forced to send thousands of letters to the land proprietors in order to negotiate the temporary occupation (. . .) If the inspection works were to start, there will be the land owners, the close family relatives and the far ones. "It will be like a river of people with the right to gather there at any moment, regardless of any police or military forces."

AAM Terra Nuova, May 12, 2008
(http://www.aamterranuova.it/article2165.htm):

> A renovated invitation from the NOTAV committees of the Susa Valley to subscribe to a share of the land: it is the strategy for opposing the construction of the new high speed Turin-Lyon rail and the related

tunnel. After Chiomonte it is Venaus's turn, for an operation to be completed by June 15 but which requires bookings by May 15. It is about the acquisition of a share of the land in the area earmarked for the building sites, in which inspections and constructions of the new infrastructure will be carried out. The share can be bought for 15 Euros and it corresponds to about 1 square metre of land.

The promoters of the nonviolent doctrine are conscious of the fact that any condemnation of violence as an instrument for political struggle is at risk of becoming an exercise of sterile morality if it is not accompanied by a serious proposal of alternative institutions and means of opposition. From here, their proposition of *satyagraha*, or positive nonviolent struggle, funded on the double thesis of a) practicability also at the level of the mass and in situations of acute conflict; and b) effectiveness as a tool for struggle aimed at the fulfilment of a society founded on the dignity of the person, the well-being of all and the safeguard of the environment (Pontara 1996).

TOWARD A POST-NORMAL AND NONVIOLENT APPROACH?

The TAV case exemplifies the complex weave of worldviews, strategies, points of view, needs that have developed over the years around a complex and controversial socioenvironmental issue. As we have seen, it is not only the start up or the blockage of a new rail line that is at stake: it is about a clash of cultures, worldviews and ideas about the nature of technoscience and its objectives; on the role of citizens and the idea of democracy. The TAV issue, therefore, is not only confined to the data that may be gathered and compared about the different aspects of the project (costs, jobs opportunities, environmental damage, health issues, freight transport etc.), but it poses important questions about the future scenarios—imagined, desired or feared ones. In order to tackle the issue, it is important to dig deeper and bring the personal and collective conscience to an awareness of the profound structures of society and culture that are mainly unconscious. For a genuine participation in matters of public life, and for a participatory democracy in which each person can express oneself and being listened to it is not sufficient to simply "instruct" the people: it is important to be able to value them, help them to express one's own individuality, legitimize them and create opportunities for encounters and for dialog. How?

> The ideal of public discussion is strictly connected with two social practices which need special consideration: tolerance of different points of view, together with the possibility of being in agreement or not, and encouragement of public discussion, together with the acknowledgment of the value of learning from others (Sen 2004, 21).

Also: "The practice of democracy offers citizens the opportunity to learn from each other, and it gives society the possibility to shape its own values and define its own priorities" (Sen 2004, 63).

Members of IRIS have recently developed and published a role-play (Camino et al., 2008b) as an educational tool to be offered in schools and universities, with the aim of promoting competences in conflict transformation, dealing with controversial issues and practicing participatory democracy.

The second part of the book will be dedicated to the illustration of numerous strategies and formative opportunities that have been inspired by these kinds of considerations. More specifically, in Chapter 8, the reflections on role-plays will show how—in our opinion—schools can help youngsters develop competences for dealing with the real-life problems we are increasingly faced with by means of approaches that are reflective, critical, conscious of complexity and nonviolent.

NOTES

1. The growing adversarial nature of this relationship is highlighted not only by the experiences of people involved in local controversies, but also by the measures and descriptions of the global state of the planet. These testify to the growing inequities among human populations in the availability of natural resources that are essential for life, as indicated by Buckles (1999); Gadgil and Guha (1995); Klare (2005); Smil (2003); Turner and Fisher (2008) and Wackernagel et al. (2002), among many.
2. In Italian: *Linea ferroviaria ad alta velocità* (TAV).
3. In Italian: *Siti di Interesse Comunitario* (SIC).
4. The Pianura Padana is the plain across the north of Italy near Milan connecting the west to the east. Traffic crossing this plain from west to east is likely to have passed through Val di Susa.
5. http://www.regione.piemonte.it/piemonteinforma/scenari/2007/giugno/tgv.htm
6. See, for example, the website www.regione.piemonte.it/piemonteinforma/scenari/2007/
7. *Carabinieri* is the Italian military police.
8. Civic disobedience, resistance and people's nonviolent defense: the lesson of the Susa Valley (Disobbedienza civile, resistenza e difesa popolare nonviolenta: la lezione della Valle di Susa), found on http://nonviolenti.org/content/view/440/2/

BIBLIOGRAPHY

Bettini, V., and C. Cancelli. 1997. *Alta velocità. Valutazione economica, tecnologica e ambientale del progetto*. Torino, Italy: Ed. CUEN-Ecologia.
Bobbio, L. 2006. Discutibile e indiscussa. L'Alta Velocità alla prova della democrazia. *Il Mulino* 1: 124–32.
Boitani, A., and M. Ponti. 2006. Infrastrutture e politica dei trasporti. *Il Mulino* 1: 102–12.
Buckles, D., ed. 1999. *Cultivating peace: Conflict and collaboration in natural resource management*. Ottawa: International Development Research Centre.

Camino, E., and A. Dogliotti, eds. 2004. *Il conflitto: rischio e opportunità*. Torre dei Nolfi (AQ): Edizioni Qualevita.

Camino, E., A. Dogliotti, B. Battaglia, G. Salio, and A. Benessia. 2008a. Problemi ambientali complessi e controversi e molteplicità di piani del contendere. Fatti, narrazioni e metafore nella questione TAV / TAC in Valsusa. In *Il dialogo tra le culture. Diversità e conflitti come risorse di pace*. Roma: Donzelli Editore, 379–414.

Camino, E., A. Dogliotti, C. Calcagno, and L. Colucci-Gray. 2008b. *Discordie in gioco. Capire e affrontare i conflitti ambientali*. Molfetta, Italy: La Meridiana Editore.

Campana, P., F. Dallago, and M. Roccato. 2006. La Tav e le grandi opere nella percezione dell'opinione pubblica (settembre-ottobre 2006) Osservatorio del Nord Ovest, Valsusa, Italia. Available online at www.nordovest.org

Cheli, E. 2002. Verso un approccio olistico al tema dei conflitti. In *La comunicazione come antidoto ai conflitti*, Convegno interdisciplinare, Arezzo, 17–18.Della Porta, D.,and G. Piazza. 2008. *Le ragioni del no. Le campagne contro la tav in Val di Susa e il Ponte sullo Stretto*. Milano: Feltrinelli.

Funtowicz, S., and J. Ravetz. 1993. Science for the post-normal age. *Futures* 31 (7): 735–55.

———. 1994. Emergent complex systems. *Futures* 26 (6): 568–82.

———. 1999. Post normal science: an insight now maturing. *Futures* 31 (7): 641–6.

Gadgil, M., and R. Guha. 1995. *Ecology and equity*. London: Routledge.

Gallino, L. 2005. Dove sta l'interesse nazionale, *La Repubblica,* December 07.

———.2006. Domande senza risposta e interesse nazionale. In *Travolti dall'alta voracità*, ed. C. Cancelli, G. Sergi, and M. Zucchetti, 39–43. Roma: ODRADEK.

Gallopín, G.C., S. Funtowicz, M. O'Connor, and J. Ravetz. 2001. Science for the twenty-first century: from social contract to the scientific core. *International Social Science Journal* 53 (168): 219–29.

Galtung, J. 1996. *Peace by peaceful means: Peace and conflict, development and civilization*. London: SAGE.

———. 2000. *La trasformazione nonviolenta dei conflitti. Il metodo Transcend: andare oltre il conflitto*. Torino, Italy: EGA.

Klare, M. 2005. *L'impero del petrolio. Internazionale* 591: 28–35.

Latour, B. 2005. From realpolitik to dingpolitik or how to make things public. In *Making things public. Atmospheres of democracy*, ed. B. Latour and P. Weibel. Cambridge, MA: MIT Press.

Mercalli, L., and C. Sasso. 2004. *Le mucche non mangiano cemento—Viaggio tra gli ultimi pastori di Valsusa e l'avanzata del calcestruzzo*. Torino, Italy: Società Meteorologica Subalpina.

Patfoort, P. 2006. *Difendersi senza aggredire. La potenza della nonviolenza*. Torino, Italy: Edizioni Gruppo Abele.

Pontara, G., ed. 1973. Initial essay to *M.K. Gandhi: Teoria e pratica della nonviolenza*. Torino, Italy: Einaudi.

———. 1996. *La personalità nonviolenta*. Torino, Italy: Edizioni Gruppo Abele.

Ravetz, R. J. 2006. Post-normal science and the complexity of transitions towards sustainability. *Ecological Complexity* 3(4): 275–84.

Salio, G. 2006. Disobbedienza civile e resistenza nonviolenta nella Valle di Susa. In *Travolti dall'alta voracità,* ed. C. Cancelli, G. Sergi, and M. Zucchetti, 45–76. Roma: ODRADEK.

Sen, A. 2004. *La democrazia degli altri*. Milan: Mondatori.

Smil, V. 2003. *Energy at the crossroad*. London: MIT Press.

Turner, R.K., and B. Fisher. 2008. To the rich man the spoil. *Nature* 451(7182): 1067–8.

Wackernagel, M., N.B. Schulz, D. Deumling, A. Callejas Linares, M. Jenkins, V. Kapos, C. Monfreda, J. Loh, N. Myers, R. Norgaard and J. Randers. 2002. Tracking the ecological overshoot of the human economy. *PNAS* 99(14): 9266–71.

4 Language and Science

Elena Camino and Martin Dodman
with Alice Benessia

EDITORS' INTRODUCTORY NOTES

Earlier chapters have highlighted the changing relationships between science and society and the implications that this has for both the nature of knowledge production and the nature of the political process. This chapter takes a closer look at language as a central dimension in the complex relationships between science, society, technology and the environment. The authors argue that language is generally considered as a mere instrument that can be used for passing information between speakers and that can be subjected to good or bad usage; yet, this conception does not take into account the dynamic nature of language, its constant evolution and the ways in which any language can be both enabling and constricting.

This chapter suggests that reflecting on language enables the development of new insights, not only in terms of the knowledge content that we build, but also about the assumptions and implications concerning the nature of science itself, and the views that we have elaborated of the relationships between ourselves and the world. The authors call for deep changes to take place in approaches to teaching and learning science in order to create awareness of the interlinked transformations in ways of living, ideas and ways of practicing and communicating science, and language. In this regard, a final reflection on visual language and its use in science communication is reported at the end of the chapter to show that the epistemological dichotomy between traditional science "speaking truth to power" and the science of uncertainty and humility can also be found in the use of images. So awareness of the role played by visual and verbal communications in conveying implicit and value-laden messages is an important element of personal emancipation, as well as a tool for an education that aims at developing young people's critical and autonomous thinking.

KNOWLEDGE AND LANGUAGE

We are witnessing a massive increase in the knowledge of our planet, but this has come with an increasing separation of most of humanity from the natural systems that support us and of which we are a part.

New aspects of reality are explored, described and explained through the development of new words or through new uses of existing words and the new concepts that can be expressed.

However, while new knowledge is widely available, there is also a widespread sense of fragmentation and loss of direction. The certainties about the world provided by traditional science—communicated through a clear, concise language shared by the experts—have given way to a multiplicity of interpretations expressed via a language that is ambivalent and often contradictory. These developments have profound implications for science education, and we believe that changes in approaches to teaching and learning science are needed in order to create awareness of the interlinked transformations in ways of living and ideas of science and language.

THE FUNCTIONS OF LANGUAGE

Language is first and foremost a way of being in the world, an instrument that enables us to conceive, organize and represent a view of reality and to act according to that view. It is a system consisting of *signifiers* (words, symbols, images etc.) that function as interconnected elements in the creation of *signifieds* (meanings). However, we do not just use a language; we use a plurality of languages. This means that we are always doing things through a variety of forms of language, performing the functions that language has evolved to permit. Developing skills in using a plurality of languages within a multiplicity of forms of communication is essential for the development of a number of cognitive and operational skills. These are knowledge-building skills, such as recognizing characteristics, relationships and transformations, understanding and interpreting processes and phenomena, constructing concepts, etc., as well as methodological and operational skills: analyzing and evaluating data, formulating hypotheses, conducting experiments, using instruments, solving problems, etc., which are necessary for the building and application of that knowledge.

We can identify three principal functions that language enables us to perform: the ideational, the interpersonal and the textual function (Halliday 1973).

The Ideational

The ideational function is the way in which language enables us to represent our perceptions and experiences, build the mental *schemata* on which we base our idea of reality and interpret what goes on within us and around us. This is language as reflection. It is language used to express what we perceive and represent as *states* (that continuously *are*) or *events* (that periodically *happen*), arranged in time and space, permanent or temporary, complete or incomplete, connected or independent, single or repetitive, fact or opinion, certain, probable, possible or impossible, cause or consequence and many other perspectives that constantly evolve depending on the uses we make of the signifiers at our disposal on the basis of our experience.

Within natural languages, different types of words (nouns, adjectives and pronouns, verbs and adverbs, prepositions and conjunctions)—and forms of these words (for example, concrete or abstract nouns or tense, aspect, mood and voice of verbs) are all used in potentially infinite combinations to express these perspectives. In English, for example, verbs used with simple aspect suggest permanence and/or completion, often implying the idea of a product, something more definitive, whereas the progressive aspect suggests temporariness or incompletion, thereby implying the idea of a process, something more subject to change. Other languages contain other forms, not necessarily verbal, used to express the same meanings. All Indo-European languages tend to use the indicative mood to express what is considered factual or "objective" knowledge, therefore definitive and permanent. Moods like subjunctive or conditional, or other forms of modality, like modal verbs, are used to express hypothesis, uncertainty, "subjective" knowledge, therefore exploratory and provisional. Voice is used to thematize one element as opposed to another. As we shall examine in more detail later, the use of nouns as opposed to verbs is often a most important way of expressing different perspectives on the reality described.

If we consider the following fairly simple examples, we can see many of these features of the ideational function at work.

- *human beings exploit natural resources* (= a fact, permanent and definitive—human beings are thematized as the agent).
- *human beings are exploiting natural resources* (= something that is happening, but potentially open to change—human beings are thematized as the agent).
- *natural resources are exploited by human beings* (= a fact, permanent and definitive—natural resources are thematized to emphasize their importance).
- *natural resources are being exploited by human beings* (= something that is happening, but potentially open to change—natural resources are thematized to emphasize their importance).
- *the exploitation of natural resources* (= a fact, permanent and definitive, rendered abstract and "objective" by the absence of an agent).

The Interpersonal

The interpersonal function is the way in which we interact with other human beings, communicate, exchange information, thoughts, memories and desires endeavor to influence, impose and so on. This is language as action. It is the basis of the way in which we act in the world in general and in the specific environments we inhabit. It is the process through which we negotiate and build meanings and thereby construct and develop our own *selves*, our ways of thinking and being. These ways of thinking and being thus derive from the systems of signifiers and meanings in which we habitually, move the forms of communication typical of the environments we frequent. Paraphrasing Wittgenstein, we may say that the limits of our signifiers and meanings are the limits of our world. Moreover, they are the result of the limits imposed by our interactions with other people, interactions in which signifiers and meanings come into contact in the minds of the participants, and in which language is both the instrument and the product of thinking and communicating: two complementary functions of any human semiotic.

The Textual

The textual function is the way in which we use language to organize signifiers and meanings and create a texture in which each element is interrelated within a cohesive and coherent whole. This weaving of connections is fundamental for the exercise of the ideational function—language as intramental reflection, forming ideas about the world—and the interpersonal function—language as intermental communication, acting within the world. Texts can be spoken and/or written and are constantly constructed, deconstructed and reconstructed when we speak and write, listen, look and read. The basis of their texture is the way in which they combine elements that are *given* (previously encountered within the text or within one's experience of the world) and *new* (encountered for the first time). This texture is the basis of all learning, a process in which we create connections between what is given and known and what is new, assimilating new meanings to given ones and accommodating, or modifying, the given on the basis of the new elements assimilated. Learning is thus the result of the interaction between the texture of ones own texts and that of the texts produced by others; the complex of intra- and intertextual connections between signifiers and meanings on which we base our lives.

LANGUAGE AS AN OPEN SYSTEM

Like all open systems, language is dynamic. Therefore, it evolves and it is *metastable*, in that it maintains itself and performs the functions for which it exists through a process of continuous change, based on exchanges with the environment of which it is a part. The evolution of language is a consequence of how we use it in thinking and in doing, in reflecting on the world and acting in it. This process can be analyzed at three intersecting levels: phylogenetic, ontogenetic and microgenetic.

The phylogenetic level concerns the development of various forms of language and individual language systems, the growth of phenomena like speech and writing or of different technologies for the production and transmission of language. The ontogenetic level concerns the acquisition and use of language by individual human beings: the move from protolanguage to the mother tongue, the development of multilingualism, the encounter with written language and various technologies. The microgenetic level refers to the single communicative events that are realized daily, specific linguistic acts that contribute to both the ontogenetic and phylogenetic development of language. The interaction between scientists and public, teachers and learners, people and the resources and materials they have access to, produces a vast number of microgenetic acts in which the role of language in learning processes is of fundamental importance.

At the phylogenetic level, every language offers a particular way of creating the texture of meanings by which its users live. All languages—including the more than 3000 natural languages, or language systems, that exist today—are solutions to the problem of how to make sense of the world around us and within us. Thus, every language is characterized by relativity, in that it opens certain horizons that at the same time inevitably pose limits to that which its users are able to conceive and express. The horizons and limits can be very different between one language and another. Moreover, the ways in which languages evolve continuously produce examples of increase or decrease of *codifiability*—the possibility of expressing certain meanings via particular signifiers—and thus a perpetual process of expansion or restriction of the language itself and its meaning potential.

At the phylogenetic, ontogenetic and microgenetic levels, all meanings are constructed through two kinds of relationships: that between signifiers and signifieds or that between signifiers, or between signifieds. In both cases, signifiers and signifieds define themselves "negatively" in terms of how they differ from other signifiers and signifieds. A given meaning is what it is because of how it differs from other meanings.

All the elements combine and recombine, creating new order, through processes of definition and redefinition of signs; of reorganization of sense. Signifiers and signifieds must remain flexible and be able to change so that the processes of creating new meaning can take place. Something new is perceived as significant/meaningful only if it can be connected to the system of signs in which we habitually move. New meanings do not cancel previous ones, they modify them.

In language, signifier and signified define each other reciprocally. One cannot exist without the other. Out of this reciprocal relationship comes the sign, the sense that we make of or give to the reality around us and within us. And every signifier continuously redefines itself through its dialog with the signified. All that we construct comes from the system of signs at our disposal, the way in which the dialog consolidates or problematizes existing signs, thereby enriching the system with new meanings that permit the construction of new knowledge, giving rise to new signs.

If, however, signifier and signified become permanently "stuck together" in the mind, the result is that word and object are (con)fused, as if they were the same thing, with a consequent loss of understanding both of the role of the signifier in the construction of meaning and of the possibility of a plurality of meanings and points of view. "In daily life we take a lot for granted, and use language in a labelling way all the time. [. . .] I call the approach 'labelling' because it seems that the word corresponds in a simple way to well-defined things, substances or actions" (Sutton 1992, 50). Changing this perspective of simplistic correspondence in order to construct awareness of the role of language in knowledge-building processes must be a primary objective for science education.

Spoken and Written Language

Language has always been primarily spoken. Written language is a very recent introduction in the history of humanity, dating from the moment in which human beings moved from a nomadic to a stable community way of life. This change brought about developments in the use of language for which speech alone was no longer sufficient. Writing provided a form of permanence in the texts produced required by such social, economic and cultural changes. It permitted the registration and recall of experience and knowledge in such a way as to render them available for consultation whenever necessary. In this way, in the words of Bruner, "writing [became] a technology for the empowerment of mind," in that it enabled the user to go back over given contents again and again and reflect (*re-flectere*), not only on the contents themselves, but also on ones own thought processes, think about one's own thinking (Bruner 1966).

Learning depends on being able to use both speech and writing for the complementary roles for which they have evolved. Speech is more immediate because of its use of the phonic channel, the fact that it happens in real time and that the interlocutors are generally present or able to interact. It is more concrete and dynamic, using a language with more verbs and adverbs and, therefore, agents, actions and results, or processes. On the other hand, writing is more abstract because it represents the sounds of speech through symbols, or graphemes, and the interlocutors are absent; therefore, meaning is exclusively realized through the lexical and grammatical resources of the language system. It is more abstract and synoptic, using a language with more nouns and adjectives and, therefore, things, facts, data or products.

Thus, a written text is a static object and imposes a synoptic view of reality, whereas speech emphasizes the dynamic nature of experience. The distinction is not merely that between alternative ways of expression, but rather that of fundamental differences in perception and representation. At the phylogenetic level, Goody (2000) describes how:

the limitations of memory in oral cultures, the role of forgetting, and the generative use of language and gesture mean that human diversity is in a state of continuous creation, often cyclical rather than cumulative, even in the simplest of societies (46).

Writing has a stabilizing function, and written cultures more easily establish norms that encompass large areas, often limiting, if not totally impeding, the spread of diversity. At the ontogenetic level, the same process occurs, described by Vygotsky (1962), as the passage from concepts that are spontaneous to concepts that are scientific. Spontaneous concepts develop through everyday experience and spoken interaction, whereas scientific concepts are the result of schooling and the encounter with a language and texts that are typically written, containing characteristic definitions, explanations, expositions and argumentations. Bruner expresses the same distinction between two modes of thinking: the narrative and the paradigmatic (1966). The narrative mode is primary, in that it is based on everyday experience and the language of doing, happening, acting and intending. The paradigmatic mode, based on written language, is a re-codification of that experience through nominal language that furnishes a particular way of managing complexity and variety and permits a reconstruction in terms of scientific concepts linked within taxonomies and subject to the operations of logic and reasoning.

Goody argues that the technology of writing not only changes the way in which we communicate, but also the very nature of what we communicate (2000). Goody identifies three levels at which this technology operates: the materials used for writing; writing as a form of storage of what has been produced; and writing as the product of the interaction between mind and the written word, which can be externalized in a way which speech does not allow. These levels can be described in terms of "cognitive procedures" an example of which is listing. Creating a list means extrapolating certain words from a flow of discourse in which they are normally interrelated and considering them as things in themselves.

> The establishment of bounded semantic categories (. . .) is a characteristic of written language, and one that places systems of classification and categorization on a different footing than is usual with the folk system of oral societies, where the categories have to be elicited and cannot usually just be read off (Goody 2000, 146).

Lists lead directly to tables,

> which are essential devices for much analytic discourse of a written kind and are impossible to conceive without writing; again, they provide a different set of dimensions for the process of reading off analogies and polarities, columns and rows (. . .) They have the effect of

forcing thought into binary forms, even when these may not be appropriate (Goody 2000, 146).

Moreover, another significant cognitive procedure made possible by writing is the kind of formalization characteristic of logic, together with the processes of reorganization of information and reformulation required by argumentation, the development of methods of demonstration, the posing of questions. Writing presupposes a process of reflection, thoughts that interact with each other and a sheet of paper, making it easier to separate things and pose questions, also to oneself.

Verbal and Nominal Language

Thus, we can see how different languages can be powerful instruments for elaborating, understanding and communicating, but also a source of misunderstanding if we are unaware of the ways in which language mediates our experience and the idea of reality that we build and share.

In particular, nominal language has played a particularly important role in the development of a scientific worldview up to the twentieth century. Scientific discourse tends to transform specific actions explicitly performed by agents into impersonal products, happening without the actions or responsibility of anyone. In this way, *people who experiment and discover* become *experimental evidence,* or *temperatures that rise and fall* become *temperature range.* The world becomes an abstraction, far from everyday, concrete experience.

Nominal language creates meanings that are synoptic, economic and "objective," typical of the current, widespread scientific view. Acquiring such language is therefore a prerequisite for learning and progressing successfully in an increasingly technological and science-based society. Yet, its acquisition is problematic in many ways. Nominal language is difficult to master because of its very nature: its lexical density (up to twenty words may be required to verbally define the complex of meanings expressed by one nominal signifier) means that its comprehension requires the simultaneous putting together of a multiplicity of meanings. Inability to handle such lexical density can lead to frustration and alienation. The consequence is that of a fracture or gap between those who possess this language and thus the power that goes with it, and those who are excluded, a division that, according to Halliday, is certainly dysfunctional in a modern democratic society (1985).

The increasing use of nominal language within our society also poses a problem because of the consequences of transforming processes, where subjects perform actions within given results, into products—impersonal events occurring without explicit causal agents. For example, to speak of *desertification* means to create a synoptic, objective abstraction that eliminates people, their actions and the consequences of these actions at three levels: the people who suffer and struggle to survive living in such conditions; the people who act and cause or exacerbate the phenomenon; and the people who might wish or be able to act in order to do something to change the situation.

In this way, the use of nominal language implies a tendency to depersonalize phenomena, de-democratize processes and deprive people of their responsibility. If, in a democracy, it is dysfunctional to be alienated by an inability to acquire a certain type of language, it is perhaps even more dysfunctional to acquire it too well! It is surely one of the most dangerous paradoxes of the phylogenesis of our language—and therefore of the ontogenesis of each one of us—that if we acquire and universally use the nominal language we have developed, then we risk being excluded or eliminated by the very same language. If we consider the following examples taken from recent newspaper headlines and captions, a type of language omnipresent in our daily lives, we can clearly see how nominal language, together with active voice, creates and spreads an idea of products rather than processes, abstract phenomena rather than dynamic events that happen or something definitive and "objective" rather than possible and contingent upon certain conditions.

- *Global warming will unleash conflicts.*
- *Climate change will cause dangerous falls in agricultural production and exacerbate tensions in already-unstable countries.*
- *Ever-greater drought and frequency of typhoons will increase destabilizing migratory pressure on the Western World.*

Within the phenomena described, human beings as agents are entirely absent, all reference to people being relegated to synoptic abstractions such as *conflicts, tensions* and *migratory pressure*. The agents are equally abstract, in that *global warming, climate change* and *drought* or *typhoons* become responsible for the consequences suffered by humanity, rather than results produced by the agency of human beings.

Critical Aspects of Scientific Language

The interdependence between language and thought has profound implications for knowledge-building processes. The gradual spread of nominal language and of interpretative categories based on comparison, extrapolation, categorization and quantification, together with the presupposition of objectivity in the description of processes, phenomena and events within the natural world, has given rise to a particular form of knowledge—defined as scientific. In recent centuries, this has assumed a privileged status with respect to other forms of thought. On the one hand, through its very language, science has been presented and perceived as knowledge that is neutral, *super partes* and free from value bias, which describes, organizes and explains facts. On the other hand, "[. . .] when I am most steeped in labelling uses, language seems like just a commentary upon nature, and not like a means of deciding what nature is" (Sutton 1992, 52).

Moreover, scientific praxis presupposes an objective and rational researcher who assumes a distanced approach to the exploration of reality. This is considered so important that in scientific publications, impersonal and nominal language prevails over narrative first-person report, and it may occur that such language hides the agents and the actions for which they are responsible. For those who do not participate directly in this scientific adventure it is difficult to imagine a space for oneself as an agent. What is presented is not a world made of processes and events in which people act and produce consequences, but rather an assembly of objects, which only experts can decipher and intervene in.

The ability attributed to scientists to offer explanations of the "material world," perceived as external to us, made of separate parts—duly labeled—leads to the role that has been assigned to them for a long time, of "teaching truth to power, deducing correct policies from incontrovertible facts" (Ravetz 2003, 64). Knowledge of facts and a position of power in decision-making processes have become crucial elements in the development of *technoscience*. Knowledge-building has lost its original speculative and descriptive role and become a powerful tool for action and transformation of natural and social systems, with consequences that should be perceived as highly controversial.

LOSS OF NATURE, EMPTIED CONCEPTS

More than half of humanity now lives in urban environments, within a context that has been deeply modified thanks to technology, with a consequent loss of contact with the natural systems on which we depend. Nature, once close and related to us, mysterious, changing and various, is now distanced from us, fragmented in a gallery of objects, labeled and defined, unable to give rise to interior resonances. We are increasingly forced to look to external sources—illustrations, photographs and descriptions—in order to produce mental images when we hear or read words of animals, plants and natural contexts.

For a vast number of people, many words are no more able to evoke the complexity of events, processes, relations and memories that match such words in the real world. Such loss of resonance may lead to indifference for a world rendered uniform by the very symbols used to describe it. These are symbols that have lost their power to evoke a personal experience through a variety of forms, odors and sizes, as well as circumstances, people, places and emotions, or capable of being experienced beyond the symbols themselves (Ferrarotti 1998). Moreover, the tendency to consider and value only what is named produces a progressive blindness toward everything else. The connections and relationships are lost and only the fragments highlighted by the labels remain.

The development of science has involuntarily aggravated the problem, owing to the enormous increase of knowledge of processes and events that are far from our direct experience. Scientific boundaries are driven ever further from those familiar to us in the spatio-temporal domains, thanks to exosomatic tools (forms of technology often extremely sophisticated and powerful) that investigate aspects and processes inaccessible to our senses and "translate" them into linguistic and iconic forms in order to make them understandable to us.

Molecular structures and black holes, extinct species and extremophile bacteria all add to an ever-increasing multitude of names, descriptions, explanations and models no longer accompanied by the richness, depth and variety of personal experience and an adequate context. Thus, the list of things we "know" (to which we give a name) increases: DNA, membrane channels, trophic networks, ecological niches, greenhouse gases, viruses, oceanic dorsals and so forth.

Complex, dynamic, transitory and unpredictable life processes give way to the "material world," a gigantic jigsaw or meccano game made up of pieces we can break up and put together again and thereby understand, control, govern.

So, urban life and specialized scientific knowledge tend to separate us—physically and cognitively—from the natural systems that contain and include us, and more and more we feel we are becoming strangers in our own world. Yet a direct, intimate relationship with natural and social systems is necessary for building knowledge and acting in the world. How can such a relationship be developed and nourished—notwithstanding the opposite trend—in such a way as to engender a sense of intimacy and continuity between us and the world?

TOWARD A SCIENCE OF COMPLEXITY

Along with the sense of fragmentation and alienation that people perceive, some discomfort is expressed within the scientific community, concerning how to deal with the increasing complexity that emerges from any field of study. As Marcello Cini puts it:

> as we rise to higher levels of organisation of matter, we find a multiplication of the languages used by different groups within a given disciplinary community based on different models of the domain of the phenomena considered and points of view (cultural, epistemological, technological) adopted for exploring it (2006, 52).

The multiplication of areas of scientific research, the proliferation of technological and conceptual tools for exploring natural systems and the increasing controversies between interpretations furnished by different schools of thought have all contributed to questioning the idea of science as an enterprise aimed at achieving a complete and objective description of the world. Some scientists argue that the recognition of the complexity of the systems implies, for science, a role of ever-greater multiplication of perspectives.

> Science does its job all too well. The argument, in brief, is this: nature itself—the reality out there—is sufficiently rich and complex to support a science enterprise of enormous methodological, disciplinary, and institutional diversity. I will argue that science, in doing its job well, presents this richness, through a proliferation of facts assembled via a variety of disciplinary lenses, in ways that can legitimately support, and are causally indistinguishable from, a range of competing, value-based political positions (Sarewitz 2004, 386).

The multiplicity of explanations and interpretations grows together with the recognition that:

> different disciplines generate different perspectives on the surrounding world, but it is too seldom acknowledged that disciplinary orientations embody different normative assumptions and divergent notions of good and relevant science. Hence, an environmental problem both looks and is inherently different when analysed from an ecological, chemical or economic perspective (Lövbrand and Öberg 2005, 196).

For some, complexity implies recognition of an irreducible plurality of relevant analytic perspectives and thus requires the endeavor to work simultaneously with numerous points of view (Gallopin and Vessuri 2006). The consequences of this are multiple and involve numerous questions, ranging from the epistemological (the nature of knowledge) to the pragmatic (the necessity and opportunity to compare and integrate interpretative schemata elaborated by different disciplines) and crucial power relationships between the "holders" of knowledge, public decision makers and society at large.

During the last twenty years, the question of complexity has become increasingly important as a result of how numerous environmental studies have contributed to a growing awareness of the changes caused by human activity on global natural systems and services and the short- and long-term consequences. Considerations of how to deal with complex systems are increasingly frequently linked to ideas, proposals and discussions concerning "sustainability." Many believe it necessary to develop new "knowledge systems," which will permit "boundary work" at the interface between communities of experts and communities of decision makers (Cash et al. 2003). A recent editorial by Clark (2007) in the *Proceedings of the National Academy of Science* (PNAS) proposes that sustainability science is not so much a field defined by the problems it addresses, but rather by the disciplines it employs. This not only has profound implications for science technology and society relationships, but it also requires fundamental advances in our conceptualization and understanding of coupled human–environment systems at the conceptual level in an effort to provide useful knowledge.

NEW FORMS OF LANGUAGE AND WAYS OF THINKING

The changes in approaches, objectives and perspectives of scientific research both involve and are profoundly influenced by language. The supremacy of the analytical and quantitative approach is questioned and elements such as narrative modes, iconic representations and other forms of creative expression assume an important role (see Benessia in Chapter 1 and at the end of this chapter, on pages 88–94). Awareness gradually emerges of how the products of science are influenced by value systems and depend on specific contexts. Alongside nominal, synoptic and impersonal language appear verbal forms that identify agents and describe processes. The role of dialog becomes crucial at various levels: interdisciplinary activities, it is generally recognized, involve people from different disciplinary areas working in an interactive way toward a common purpose. As the disciplines involved in such interdisciplinary activities are, in most cases, scientific specializations or professional areas, the participants will tend to share some kind of basic platform of beliefs, e.g., trust in the scientific method. However, considerably greater challenges are presented, for both research and action, by interparadigmatic activities. It cannot be assumed that a common platform of beliefs will be held, or can be imposed, and there may not be a sense of common purpose. In such situations, it is not just how to reconcile different worldviews, but also different and equally legitimate goals. It will not be possible, or indeed desirable, to reduce the plurality of viewpoints and interests to a single format, e.g., a mathematical model or a narrative representation (Modvar and Gallopin 2004).

The multiplication of knowledge—a result of an increasing awareness of complexity—has a number of interesting implications. A first example is the direct link that is revealed between knowledge-building and increasing ignorance: "additional research reveals heretofore unknown complexities in natural systems, or highlights the differences between competing disciplinary perspectives, and thus expands the realm of what is known to be unknown" (Sarewitz 2004, 396).

A second consequence is the focus on the transitory, provisional and relative nature of knowledge. According to Tattersal (2008), the process of science is much like negotiating a hugely complex maze with false starts and backtracking. But, as he points out, it is no crime to be wrong in science and scientists should not be out to *prove* anything, it is a mistake to confuse quantifiability with objectivity. "[. . .] where did the ideas, good or bad, come from in the first place? Well, there are no rules for human creativity—how could there be?—and science depends on creative thought and intuition quite as much as any other branch of human endeavor" (Tattersal

2008, 39). "The best science can offer is a temporary consensus on how to interpret certain evidence" (Lövbrand and Öberg 2005, 196).

A third consequence is that the boundary between expert and nonexpert becomes blurred. Gallopin et al. (2001) argue that when confronted by the complexity of the problems that modern societies have to face, the most rational approach would be to increase as much as possible the different perspectives employed. Many people whom academics would consider "nonexperts" are able to bring to bear new perspectives that are crucial for the understanding of complex problems. Particularly at the local level, the discovery is being made, again and again, that people can be ingenious and creative in finding practical, partly technological, ways toward the solution of problems: they can imagine solutions and reformulate problems in ways that the accredited experts, with the best will in the world, do not find "normal" within their professional paradigms.

On the other hand, it is by no means certain that a scientist highly specialized in one well-defined field can offer a useful contribution in another, only partially different, field. As suggested by Levy-Leblond (1992), because of the state of scientific specialization, ignorance about a particular domain of science is almost as great among scientists working in another domain as it is among ordinary members of the public. He goes on to suggest that there is no single general knowledge gap between scientists and nonscientists. Instead, there is a multitude of specific gaps between specialists and nonspecialists in each field. "Science is not a large island separated from the mainland of culture, but a vast and scattered archipelago of islets, often farther apart from one another than from the continent" (17).

Another consequence is that power relationships at the institutional level are called into question. The role of scientists in decision-making processes and governance in general are becoming controversial. The spread of knowledge—no longer confined to a closed scientific community, but open to many parts of society and different actors; no longer universally considered unique and "scientific," but rather deconstructed and reconstructed within epistemologies that are different and diversely pertinent—offers profoundly different perspectives on all those involved, their interrelationships and the question of validity and accountability (Tallacchini 2005).

A final consequence concerns the relationship between science and nature. As we have already emphasized, science drives the idea of nature—and of oneself—that people develop. Scientific language is not only a means to label newly discovered aspects of the world, but also a means to convey (often implicitly) an underlying idea of the relationship existing between nature and humans. According to Beck et al. (2003):

first modern societies have a particular *concept of nature* founded on its *exploitation*: Nature is simultaneously central to society and marginalized. It appears as the "outside" of society. Nature is conceived of as a neutral resource, which can and must be made available without limitation. Rational progress is conceived of as a process of demystification that can continue without limits. This implies a belief that *scientization* can eventually perfect the control of nature.

A different vision is held by Stephan Harding (2006), who suggests that we can act from a deeper, wider mode of consciousness in which we come to realize that Gaia is beyond our control and that it is impossible for us to ever be the masters or stewards of the Earth.

So, language—and scientific language in particular—is a powerful tool through which a picture of the relationship between ourselves and nature is continuously created and transformed: a tool that can be used to lead humans toward a more or less sustainable way of living.

PERSPECTIVES FOR SCIENCE EDUCATION

Science: From a Product to be Learned to a Process of Participation

As we have seen, language is a means of conveying an idea of science, a view of the world, a model of power relationships. Attention to language should therefore not just be a question of reflection on the part of the teacher, but it must also play an important role in educational practice and involve the gradual development of language awareness on the part of learners. In this respect, Désautels and Larochelle (1998) offer an example of a conversation between a teacher and learners in which:

> the language used by the teacher ignores not only the theoretical context which informs and gives meaning to the concepts of wave, electricity and electron but also, and most especially the deliberative activities by which scientists ultimately agree on the relevance of such concepts for solving the questions and problems they are tackling.

Actually, the entire sequence of events appears to unfold as though there were a one-to-one correspondence between the concepts under consideration and a number of real entities that could be pointed out and that scientists have simply discovered and named: here we have a wave; here we have an electron. They go on to say that there is the same tendency to picture scientific knowledge as *knowledge of something* rather than as knowledge that is socially constructed and negotiated and, in keeping with what they term as the "thingifying" vision of science, teaching strategies are made use of in which *telling* and *showing* predominate. In other words, these strategies are unlikely to grant students' experience-based knowledge any sort of relevance.

Sutton (1992) offers a further example of the importance of reflecting on language for science education. He suggests that, because labels are important products of scientific activity, it ought not to be argued that they should never be taught in school. What he does suggest is that the habitual experience of labeling encourages a set of beliefs about language that can quickly de-skill learners by cutting them off from the habit of reprocessing ideas. The argument is that if learners are to get a feel for language as an interpretative system, they must have experience of using it in that way themselves. He goes on to say that scientific language is commonly seen as simply a means of recording facts, but that we should give more prominence to is its use as an instrument of thought and its role in developing a sense of social identity for members of the scientific community. As such, for pupils in a classroom, the language of science deserves much more attention.

Increasing attention is being paid to the role of discourse and argumentation in the classroom, both as a way of promoting understanding of the nature of science and of rendering learners protagonists of their own learning. Fendler and Tuckey (2006) argue that language is not separable from science, nor is language an unfortunate but inevitable inconvenience that tends to get in the way of real science. Instead, they regard science as a discursive practice.

It is particularly interesting to reflect about the role of metaphor in science. Metaphors offer a highly effective way of defining new content in thought, resulting from the encounter with a new reality (object or situation) or a new form of relation with a given reality. Lakoff and Johnson (1999) reflect on the use of metaphor and point out that, as a result of our normal functioning in the world, we automatically and unconsciously acquire and use a vast number of metaphors. These metaphors are a consequence of the nature of our brains, our bodies and the world we inhabit. In addition, they suggest that most, if not all, of our abstract concepts are defined in significant part by conceptual metaphors. According to these authors, metaphorical thought is what makes abstract scientific theorizing possible. It is the very means by which we are able to make sense of our experience. Reason and conceptual structure are shaped by our bodies, brains and modes of functioning in the world.

Democratizing Science and Education

Awareness of the role of language in science is thus fundamental for understanding the crucial political role of science teaching in terms of facilitating or hindering a process of democratizing science. The objectives of science education, the approaches to educational practice, the language used and the relationships created between teachers and learners both reflect and are a vehicle of an idea of society and the role we attribute to ourselves and to the new generations we endeavor to educate. Within the model proposed by *post-normal science*, the observer is a subject among many, aware of having a limited perspective on any issue and aware that no perspective can be better than others. Moreover, all the persons who are involved in the problem should be able to develop a dialogical relationship with each other, and be allowed to contribute toward finding a solution.

As observed previously, scientific language can be seen as either empowering or an obstacle to participation, according to how nominal, synoptic language is mastered and employed. The challenge is that of finding ways of developing an "emancipative relationship to scientific knowledge" (Larochelle and Désautels 1991).

In the following pages, we begin to explore this idea of emancipation by drawing on experiences of research in both science education and the arts. First, we introduce the methodological aspects of an educational practice that draws on the insights illustrated here about the role and function of language in knowledge construction. Then we illustrate the links between use of images and perpetuation of views of reality, models of development and sociopolitical dynamics.

Experiences in Science and Language Education

In Chapter 6, we will sketch a few examples of activities performed during tertiary education courses in which the participants (young researchers or future teachers of science in secondary school) were asked to reflect, produce thoughts, discuss among themselves, in-line with our teaching-learning style, based on reflection, interaction and interdisciplinarity. Our approach is in tune with the aim of reconnecting ourselves—human beings—with the natural systems we are included in. Even though we will present only a few examples, we hope they will give a flavor of the kinds of processes initiated.

The "modules" that we put forward, as described in other chapters, include a variety of different activities: individual reflexive contributions, discussion in small groups, open questions within the context of the class, meta-analysis of the various perspectives that emerge from the students, epistemological and linguistic reflection, links with socioenvironmental issues etc.

The initial step is usually that of encouraging the participants to freely express their own ideas about a concept, an issue or a question by means of some sort of brainstorming. The ideas, expressed in the form of written personal reflections, answers to questions posed by a questionnaire or contributions to group brainstorming, form an outline, a frame for subsequent steps. This way of starting becomes all the more welcome and involving as familiarity grows within the group and the participants become aware that the objective of such activity is not to assess their previous knowledge, but rather to facilitate participation on the part of all and to appreciate and make use of their contributions in a process of co-construction of new, more articulate, aware and critical knowledge. In this way, the students are motivated to participate in an activity where they are the protagonists, while the exchange of ideas, which always goes well beyond the mere passing on of notions, promotes the development of open and friendly personal relationships.

People involved in such activities, whether or not they have a similar university background, produce a considerable variety of views about scientific arguments, so that this educational strategy provides a good basis for developing new attitudes to scientific learning, paying attention to the processes and not just the products of knowledge building. Moreover, future teachers who participate in various and articulated activities in which they experience directly interactive, reflective, interdisciplinary and inclusive approaches, gradually acquire methodological, operational and linguistic competences and confidence in their ability to handle similar ways of working with their students.

We invite interested readers to go to Chapter 6 for a detailed description of some specific educational activities dealing with language and science.

The following section takes us back to a reflection on language in science–society interactions, giving further details of the relationships between use of languages, views of the world and power relationships.

VISUAL LANGUAGE IN SCIENCE: FROM EXPERIMENTAL EVIDENCE TO OPEN DIALOG—ALICE BENESSIA

When talking about language, not only words but also images have played a crucial role in representing and understanding natural phenomena. Visual language used long before the birth of the scientific method have always played influential roles in the formation of meanings and messages (see, for example, the use of iconography in the religious domain), and later, they made a substantial contribution to the creation, legitimization and communication of scientific results. Scientific pictures in particular are habitually interpreted as faithful and neutral reproductions of a given portion of reality. However, all images, including scientific ones, carry factual as well as a metaphorical content. Both levels of meaning are the result of a *framing* process, both in a literal and figurative sense: a series of assumptions embedded in the observation and representation, such as the spatial and temporal scale, the instruments involved and the selection of boundaries. Moreover, observing, representing and understanding entail the extrapolation of the unknown from the known. This process involves a more or less conscious discrimination between relevant and superfluous information, which in turn depends on the observer's intellectual approach, aims and preconceptions of reality.

As with any other kind of visual products, science images are then relevant not only and not mainly for *what* they show, but also for *how* they show it and *why*, and at the same time, naturally, for what they *do not* show. Indeed, as we have seen in Chapter 1, the normative power of the master narratives that shape our collective imagination, such as the modern ideal of scientific rationality and objectivity as a privileged form of knowledge, rests on the mechanism of selecting what counts as "evidence," that is, what is not and *needs not* to be seen. So, reflecting on the phylogenesis and the microgenesis not only of natural language but also of visual representations becomes, in this context, a fundamental tool: on the one hand, for understanding the ways in which these narratives emerge and are diffused into our culture, and on the other, for cooperatively elaborating new narratives that can envision a more sustainable future.

From Experience to Experiment

The first example of human reproduction of natural phenomena in the form of mural drawings, which were both scientific and artistic images at the same time, is dated 30,000 years ago, long before the birth of written language (Aujoulat 2005). Human beings recorded the images of their world, animal figures and hunting scenes, much earlier than their words. Drawing was born as a way of comprehending reality by translating a visual act—the experience of seeing—into a physical act—the hand gesture. This translation allowed capturing the impermanence of natural phenomena in order to reflect on and understand them. It stemmed out of a search for meaning: artistic, scientific and spiritual—rational, intuitive and emotional at the same time. Knowing was deeply embedded in the unmediated experience of natural phenomena. In this holistic approach, which survived into different cultures for millennia and was embraced by the genius of Leonardo Da Vinci, the quality of the gesture corresponds to the quality of experience and to the quality of knowing. Representing, interpreting and understanding are one and the same thing (Kemp 2004; Capra 2007).

With the birth of experimental science, the function of drawing collapsed into a far more restricted function: the neutral reproduction of a given portion of reality. Through the development of the experimental method, natural phenomena were confined in controlled and simplified environments. At the same time, new exosomatic technologies, such as telescopes and microscopes, transformed the act of observation into a mediated and elitist experience. In order to communicate and legitimize their astronomical and microscopic findings, Galileo Galilei and Robert Hooke had to repeatedly affirm the faithfulness of their drawings by excluding any subjective interpretation and aesthetic concern:

> In this kind I here present to the World my imperfect Indeavours; [. . .] I hope, they may be in some measure useful to the main Design of a reformation in Philosophy, if it be only by showing, that there it not so much requir'd towards it, any strength of Imagination, or exactness of Method, or depth of Contemplation [. . .] as a sincere Hand, and a

faithful Eye, to examine, and to record, the things themselves as they appear. (Hooke 1662, 8)

From the challenge to comprehend and express the marvel of unmediated experiences of natural phenomena, the process of understanding and knowing based on observation became a matter of experimental objectivity. The quality knowledge was then identified with the conscious effort of self-effacing the subjective individuality of the investigator in order to reveal "the things themselves as they appear."

Galileo took a step further toward the ideal of objectivity by associating the process of investigating and understanding the complex world of natural phenomena with the act of decoding the "book of nature," written in the visual symbolic language of Euclidean geometry. The search for regularities, universal principles and properties of natural philosophy was then reformed, as mentioned by Hooke, and became anchored to the neutral representation of phenomena and to the use of the abstract language of geometry.

Science in the Age of Mechanical Reproduction

Around 1830, the invention of photography and of other kinds of mechanical reproduction instruments—such as kymographs, which record changes in pressure by means of a stylus marking a drum—determined a new development of the ideal of objective representation by assigning to an external device, neutral by definition, the task of recording the results of scientific investigations. Scientific evidence was then identified with the mechanical reproduction of experimental outcomes: it was the birth of "mechanical objectivity," as the historians Daston and Galison called it in a recent work (2007). The authors argue that, in this transition, objectivity became an "epistemic virtue," a fundamental moral attitude of the ones that were legitimately recognized as the makers of knowledge, based on a self-cultivated capacity of refraining from the expression of the unreliable world of subjective emotions. The Cartesian dualism between mind and matter and the identification of fair knowledge with objective and rational thinking was embodied in this approach.

By the end of nineteenth century, images lost their dominance in the scientific discourse. With the rise of physics as the leader of all sciences, and of reductionism as the dominant paradigm, the conception and use of visual language was further reduced to the illustration of mathematical concepts. Quality of knowledge was still associated with objectivity, but this latter was not ensured by mechanical reproduction, but by quantitative thinking. Daston and Galison refer to this kind of quantitative and logical neutrality, associated with the positivism of the Vienna circle, as "structural objectivity" (2007). It was the birth of what Gregory Bateson defined as the "anti-aesthetic assumption, borrowed from the emphasis which Bacon,

Locke, and Newton long ago gave to the physical sciences . . . that all phenomena (including the mental) can and shall be studied and *evaluated* in quantitative terms" (Bateson 1979, 235; emphasis in the original).

High-power Technoscience: Visual Language in the Digital Era

As illustrated in Chapter 1, the twentieth century was characterized by the relentless development of the technoscientific capacity to act upon social and natural systems. Scientific and technological research and implementation moved out of the controlled and simplified environment of the laboratories and entered into a phase of open-field experimentation over local and global scales. Concurrently, the capacity to process data was exponentially enhanced with the invention of modern computers. Numerical simulation, digital visualization and, most recently, web communication technologies, became the essential tools of technoscientific progress. As a result of this constantly growing computational power, structural objectivity began to be associated with statistical analysis and numerical simulation. At the same time, mechanical objectivity evolved into the more fluid notion of "trained judgment" (Daston and Galison 2007), intended as the capacity of the laboratory personnel to interpret the results of complex experimental apparatus and to make use of the most sophisticated visualization technologies. Upstream hybrid technosciences such as biotechnology and nanotechnology were borne out of this revolution. The quantitative approach of statistics and data management on the one side, and the visual language of both organic and inorganic forms on the other, became their fundamental tools.

As we mentioned, in conjunction with the growing power to manipulate and act upon socioenvironmental systems, contemporary technoscientific research involves uncertainty, risks, indeterminacy and ignorance. Consequently, the modern model of science as a privileged form of certain, objective and exhaustive knowledge is deeply undermined (see Chapter 1). One attempt to contain this crisis and to preserve the ideal of the "republic of science" (Merton 1968) is to appeal to the structural objectivity of statistics by confining the area of the unknown into the quantitative terms of numerical simulations, risk management and cost–benefit analysis: in other words, by utilizing what we referred to as technologies of hubris (Jasanoff 2003). But the specialized and synoptic language of statistical analysis is not easily translatable into common terms for reaching and reassuring an extended community of citizens.

A different strategy for preserving the privileged status of technoscientific knowledge into our society of risk relies on the aura of objectivity of scientific visual evidence. This kind of approach, increasingly present in our cultures and societies, is fully articulated in the work of Felice Frankel, a science photographer, research scientist at Massachusetts Institute of Technology and head of the Envisioning Science program at Harvard's

Initiative in Innovative Computing. In March 2002, Frankel published a book called *Envisioning Science: The Design and Craft of Science Images* (Frankel 2002). Her extensive and colorful volume was not addressed to the general public, but to the science community. It was, in fact, presented as a technical manual to keep in the lab in order to improve the quality of scientific images for journal submissions, funding agencies, investors and the general public. Her work was the result of a few years of intensive in-the-field collaboration with scientists and engineers to create more accessible, compelling and beautiful science pictures, some of which were covers of *Science* and *Nature* magazines. As we read in her introduction:

> This book is about a *new kind of science image*, an image that communicates your work more effectively to both colleagues and the general public. [. . .] The science pictures you see . . . have a component that is sometimes called 'artful' a word, I, like you should be wary of using. They might appear as personal interpretations but they are not. They are honest documentations of scientific investigations. However, they have an *additional quality* not usually present in science image—they somehow include the marvel of whatever phenomena I intend to capture. (1; emphasis added)

A new kind of image is proposed as a way to communicate the results of technoscientific results both inside and outside of the community of peers. In Frankel's approach, the elitist language of science has to be democratized by opening the laboratories to the unspecialized gaze of the general public. The qualitative and visual language of experimental evidence seems to be the optimal tool to achieve this aim. Indeed, in Frankel's words:

> Using compelling and accessible pictures is a powerful way to draw the public's interest to the world of research. When the public develops a more intimate association with science the results will be both a richer society and one supporting the important efforts in scientific investigation (1)

But what model of scientific investigation is communicated through the images? What kind of scientific research needs to be supported? In more general terms, what kind of narrative is involved in this conception and use of scientific visual language?

The factual content of the images is determined by the disciplinary approach, the experimental procedures and aims: it can be a 3-cm drop of ferrofluid on a glass slide, a set of semiconductor nanowires or any other experimental sample. The metaphors implied are conveyed through the "additional quality" of the images: the marvel of an honest documentation of scientific findings for what they are. The reference to Hooke's "sincere hand and faithful eye" on the one hand, and to mechanical objectivity as an epistemic virtue on the other is clear. But, in this case, honestly documenting does not exclude an aesthetic concern, although mediated through a "trained judgment." Indeed, the subjective world of emotions is still confined to the realm of artistic expression, and this latter is explicitly distinguished from the objective beauty implied into the scientific enterprise.

> [Artists] take a line that they see in the science and just expand it without any . . . I'm going to say "reverence" to the information in the science. So that's why it's very important for people not to look at my pictures as art. That's not my intention. The images are not about me or my emotions. They're images of science that I've used photographic tools to bring out. [. . .] A "pretty" picture of science is not mere decoration but one that *reveals the beauty and substance that is already there.* (Frankel 2008, interview in ScienceBlogs, www.science blogs. com; emphasis added)

Scientists have then to acquire a new form of visual expertise in order to fully reveal the objective beauty and marvel at what they discover and make in their research. The message implied in this approach is that technoscientific research uncovers and controls the marvel of natural phenomena; *therefore*, it is worth supporting. The objective beauty of the images works as evidence of the intrinsic value of the science involved. What the visual findings do not show, and what needs not be seen in the construction of this visual narrative, is the overall framing in which the specific technoscientific research is performed. The normative aspects involved, such as the aims and the issues of distribution and vulnerability (see Chapter 1), are all but explicit. Moreover, the narrative implied in the metaphoric content of this kind of imagery is the one of laboratory science, in which experimental results are certain and the phenomena involved are under perfect control. In other words, in analogy with Jasanoff's technologies of hubris (Jasanoff 2003), these kind of visual products are based on an overstatement of the known at the expenses of the unknown.

Therefore, the model of science that remains implicit in this context is again the modern ideal of a privileged and certain knowledge; the result of an objective investigation of natural phenomena in which the normative assumptions and consequences of technoscientific research *are* and *can* be excluded. There seems to be little room for an open discussion about the model itself, and these newly developed visual tools, both conceptual and technological, are essentially utilized for its preservation as sophisticated marketing devices.

Back to Experience: Visual Language in the Post-normal Approach

As we have seen, in the post-normal framework, the public engagement and the decision-making process about socioenvironmental issues are not based on a scientific demonstration of all the relevant facts, but on an open dialog between all stakeholders in which the normative dimension is and has to be explicitly taken into account. Different kinds of expert knowledge are equally appropriate and legitimate.

In this scenario, the visual language of science ceases to be interpreted and used as a privileged and objective form of evidence. Images become a platform for dialoging about framing issues, values and priorities. Both the factual and the metaphorical content are conceived of and utilized not only as descriptive, but also as projective elements. Indeed, when looking at the same image, different stakeholders see and talk about different aspects, making explicit their legitimate preconceptions and visions of the issues at stake. Climate-change visual representations and simulation models are well-established examples of this modality (Ravetz 2003; see Chapter 1).

As we have mentioned previously, democratizing scientific knowledge implies, on the one hand, the institution of proper normative structures in which an extended participation is possible, and on the other hand, the elaboration of conceptual and methodological means for stimulating public awareness, critical and creative thinking. In this context, the modern divide between the subjective world of artistic expression and the objective world of scientific representation becomes obsolete. The epistemic value of artistic investigations in terms of the visions they produce on the one side, and the normative value of scientific research in terms of the metaphors they convey on the other, have to be acknowledged as relevant and legitimate perspectives on the complex reality we act upon. In order to cooperatively envision new narratives of sustainability, a new holistic approach to natural phenomena has to be recovered, one in which the quality of knowledge emerges from the quality of experience, being it artistic, scientific or simply human.

BIBLIOGRAPHY

Aujoulat, N. 2005. *Lascaux: movement, space and time.* New York: Harry N. Abrams Inc.

Bateson, G. 1979. *Mind and nature: a necessary unity.* New York: E.P. Dutton.

Beck U., W. Bonss, and C. Lau. 2003. The theory of reflexive modernization. *Theory, Culture & Society* 20(2): 1–33.

Bruner, J.S. 1966. *Toward a theory of instruction.* New York: W.W. Norton.

Capra, F. 2007. *The science of Leonardo: Inside the mind of the great genius of the renaissance.* New York: Doubleday.

Cash, D.W., W.C. Clark, F. Alcock, N.M. Dickson, N. Eckley, D.H. Guston, J. Jager, and R.B. Mitchell. 2003. Knowledge systems for sustainable development. *PNAS* 100: 8086–91.

Cini, M. 2006. *Il supermarket di Prometeo.* Torino: Codice.

Clark, W.C. 2007. Sustainability science: A room of its own. *PNAS* 104(6): 1737–8.

Daston, L. and P. Galison. 2007. *Objectivity.* New York: Zone Books.

Désautels, J. and M. Larochelle. 1998. *About the epistemological posture of science teachers. Connecting research in physics education with teacher education.* International Commission on Physics Education. Available online at http://www.physics.ohio-state.edu/~jossem/ICPE/D3.html

Fendler ,L., and S.F. Tuckey. 2006. Whose literacy? Discursive constructions of life and objectivity. *Educational Philosophy and Theory* 38(5): 589–606.

Ferrarotti, W. 1998. Rapporto diretto con l'ambiente: condizione di significatività e validità della conoscenza. In *Proceedings of VIII Seminar of Didattica delle Scienze Naturali*, Novembre 7 and 8, Scuola Ambiente and Parchi. Centro OASI—Cavoretto (TO).

Frankel, F. 2002. *Envisioning science: The design and craft of the science image.* Cambridge, MA: MIT Press.

Funtowicz, S., and J.R. Ravetz. 1999. Post normal science: an insight now maturing. *Futures* 31(7): 641–6.

———. 2001. Post-normal science—Environmental policy under conditions of complexity. In *Interdisciplinarity in technology assessment. Implementation and its chances and limits*, ed. M. Decker, 16–24. Ethics of Science and Technology Assessment 11.

Gallopin, G., and H. Vessuri. 2006. Science for sustainable development. Articulating knowledges. In *Interfaces between science and society*, ed. A. Guimarães Pereira, S. Vaz, and S. Tognetti, 35–51. London: Greenleaf Publishing.

Gallopin, G.C., S. Funtowicz, M. O'Connor, and J. Ravetz. 2001. Science for the twenty-first century: From social contract to the scientific core. *International Social Science Journal (NWISSJ)* 53(168): 219–29.

Goody, J. 2000. *The power of the written tradition.* London: Smithsonian Institution Press.

Halliday, M.A.K. 1973. *Explorations in the functions of language.* London: Edward Arnold.

———. 1985. *Spoken and written language.* Oxford: Oxford Univ. Press.

Harding, S. 2006. *Animate Earth. Science, intuition and Gaia.* Totnes: Green Books.

Hooke, R. 1662. *Micographia.* London: Science Heritage, 1987.

Jasanoff, S. 2003. Technologies of humility: citizen participation in governing science. *Minerva* 41(3): 223–44.

Kemp, M. 2004. *Leonardo: Leonardo seen from inside out.* Oxford: Oxford Univ. Press.

Lakoff, G., and M. Johnson. 1999. *Philosophy in the flesh.* New York: Basic Books.

Larochelle, M., and J. Désautels. 1991. The epistemological turn in science education: The return of the actor. In *Research in physics learning: Theoretical issues and empirical studies*, ed. R. Duit, F. Goldberg, and H. Niedderer, 155–75. Kiel: Institute for Science Education.

Levy-Leblond, J.M. 1992. About misunderstandings. *Public Understanding of Science* 1: 17–21.

Lövbrand, E., and G. Öberg. 2005. Comment on "How science makes environmental controversies worse" by Daniel Sarewitz, *Environmental Science and Policy* 7: 385–03. and "When sientists politicise science: Making sense of the controversy over The Skeptical Environmentalist" by Roger A. Pielke Jr. *Environmental Science and Policy* 7: 405–17. *Environmental Science and Policy* 8: 195–7.

Merton, L.K. 1968. *Science and democratic social structure in social theory and social structure.* New York: Free Press.

Modvar, C., and G.C. Gallopin. 2004. Sustainable development: Epistemological challenges to science and technology. CEPAL no. 44, Santiago, Chile, October 13–15.

Ravetz, J. 2003. Models as metaphors. In *Public participation in sustainability science: A handbook*, ed. B. Kasemir, J. Jäger, C.C. Jaeger, and M.T. Gardner, 62–79. Cambridge: Cambridge Univ. Press.

Sarewitz, D. 2004. How science makes environmental controversies worse. *Environmental Science and Policy* 7: 385–403.

Sutton, C. 1992. *Words, science and learning.* Buckingam: Open Univ. Press.

Tallacchini, M. 2005. Before and beyond the precautionary principle: Epistemology of uncertainty in science and law. *Toxicology and Applied Pharmacology* 207: 645–51.

Tattersal, I. 2008. What's so special about science? *Evo. Edu Outreach* 1: 36–41.

Vygotsky, L.S. 1962. *Thought and language.* Cambridge, MA: MIT Press.

Part II

Science and Sustainability

Implications for the Learning and Teaching Process

Introduction to Part II
Towards a Sustainability Education

Donald Gray, Laura Colucci-Gray and Elena Camino

The discussions of science–technology and society issues conducted in the first part of the book have been conducive to the projection of an image of science and society in continuous and dynamic interaction: scientific enterprise—with its strengths and problematic aspects—is both a product and a process of societal life. This is a much more fluid scenario than the conventional idea of science "speaking truth to power." The nature of democratic life plays a crucial role in moderating their interaction (i.e., the use and impact of science on society). In particular, the role of scientists—as the experts—changes from the position of neutral, objective, external observers to that of involved participants in an arena where multiple perspectives and points of view are compared and shared. Conversely, the epistemological framework of post-normal science extends the ethical implications and includes them in its methodological stance, putting forward a vision of collective responsibility and participation in the governance of science related issues. Chapter 4 contributed further to this discussion by adding some observations on language: How this can be a barrier to interaction with other people and groups, as well as a means for sharing experiences and promoting mutual understanding? Particularly in relation to science, an increasing awareness of the power of language can disclose precious insights about scientific research, as well as about learning and using science. So, in the post-normal science framework, language becomes a central aspect, shaping the nature of the collective approach to knowledge and learning.

In the context so outlined, the second part of the book addresses the question of education, and particularly the central role that science education can play in preparing citizens to engage in a context of civic participation. To begin this reflection, we start with a brief review of some of the main developments that have characterized science education in recent years. We will see that education is located at a critical interface between conceptions of science and associated models of society, and much of the debate that takes place in the educational community around the notion of scientific literacy often originates from different ways of conceptualizing this (dynamic) relationship. In conducting this review of recent research, our aim is to provide the frame for the contributions that follow, and that will expand the reflection on a science education for sustainability.

SCIENCE LITERACY AND CITIZENS' PARTICIPATION IN GLOBAL ISSUES

As indicated in Chapters 1 and 2, growing concerns about sustainability raise the need for the involvement of citizens in an extended peer community. Accordingly, this widens the notion of "expertise" from being a character of single individuals holding specialized knowledge, to that of being a competence, an ability, of a collective group of people to engage in the production of new knowledge, along with new discourses, models and images (Benessia, Chapter 1). Yet, this shift implies a change in the traditional conceptions of science education and science literacy (Hodson and Prophet 1993; Hodson 1999).

The field of science–technology–society (STS) education, introduced in the early 1980s, set the beginnings of a discussion on the significance of acquiring a scientific literacy. At that time, Western societies witnessed the first explosion of socioenvironmental problems: waste disposal, water pollution and shortages in energy production began to populate the political agenda, generating uncertainty and confusion for both the public and the decision makers alike. In many northern countries, such as the United Kingdom, but also in Northern Europe and North America, science education began to respond to what were perceived as being new risks and new threats for society by introducing curricula aimed at preparing students with the knowledge, attitudes and skills for making responsible decisions (Solomon 1993). In this scenario, the acquisition of scientific literacy was aimed at enabling citizens to carry out their own inquiries by gathering their own sources of knowledge and evidence and in so doing, become "expert" citizens who were able to use scientific knowledge to assess risks and make the right decisions. It was a time when scientific literacy was seen as a tool for the public to become actively involved in consultation and scrutiny of the impact of science. However, STS education was to face at least two orders of problems. The global dimension of environmental issues required an understanding of the interconnections among science, economics and the environment on a global scale. This would require an interdisciplinary approach that was uncommon in schools. In addition, a focus on disciplinary content and notions would often take priority over more reflective aspects of science learning. So, at the end of the 1980s, a focus on factual knowledge in science was prioritized over a concern for the development of competences and engagement.

In the meantime, there were growing concerns about conventional ways of teaching and learning science that were strongly focused on the transmission of specialized, factual knowledge, with the dominance of teacher-centered approaches. Science was presented as a special form of knowledge: concerning only "facts," and also being more accountable and reliable than other forms of knowledge (Lederman 1992). In addition, within science itself there was still a tendency toward teaching separate subjects without any systematic effort at building bridges between disciplines with different epistemological positions.

In practice, the answers given by science education to STS problems were varied and, to some extent, divided on different fronts. The increasing apprehension over environmental problems called for the need to acquire an ecological literacy (Orr 1992) and to become aware of the systemic nature of socioecological systems (Capra 2002). In other contexts, such as the United Kingdom and The Netherlands, a growing sensitivity toward public engagement with science supported the discussion of socioenvironmental issues as part of the formal curriculum, with a concern for raising the interest of the pupils who had become disaffected toward science learning. Such endeavors, however, brought another set of concerns, such as whether or not students would make use of scientific ideas and concepts in seeking to settle the controversies. In some quarters, this problem was addressed with structured approaches, such as the teaching of the skills of argumentation to enable students to understand science by adopting some of its language processes (Erduran and Jimenez-Aleixandre 2008; Osborne 2003). For some authors, this approach could do little for the fluidity of ethical reasoning characterizing STS problems, "where there are likely to be many grounds for doubt which have to be faced, rather than solved" (Solomon 2004, 1895).

So, in the historical context of the 1990s, there was more recognition of the need to extend consultation and civic responsibility, yet such efforts were strongly shaped by an idea of science that did not include awareness of complexity and uncertainty. Behind this scenery also lay a propensity to divide people as "experts" and "lay persons" and an inadequate perception of the complexity of natural systems and of the outcomes and impact of one's own individual and collective action upon the planet:

> The pursuit of knowledge is not a simple observation process that simply augments the stock of knowledge about raw materials that is put on the market shelf. Rather, it is an intervention process that, through learning by doing, gives knowledge about possibilities of induced transformation (Gallopin et al. 2001, 226).

Interestingly, in other contexts, a number of educators warned about the need to value students' own ideas and epistemologies as legitimate aspects of the process of learning and to encourage dynamic interaction between the reality of the students and the reality of scientific concepts. As reported by Roth and Calabrese-Barton (2004): "Enacting scientific literacy hinges on much more than the content of the educational activities explored. It hinges on the expression of community from moment to moment" (122).

This way of teaching would also compensate for the rigid worldviews encapsulated in "thingifying" and objectifying images of scientific knowledge (Larochelle and Désautels 1991). In addition, other authors warned about the need to legitimize the knowledge of students from other cultures, to also include what is referred to as folk knowledge of first nation people (Aikenhead 2006). Such approaches would open the way for a scientific literacy that is not solely concerned with the application of scientific concepts for solving problems, but focuses on developing awareness of the conceptual infrastructures of scientific knowledge (i.e., how it functions, the way it represents the world). In practice, this meant recognizing the centrality of the students as holders and active producers of knowledge, and such an approach can also incorporate a more dynamic view of language as an opportunity for students to express themselves and to engage with different forms of literacy, each one bringing forth different ways of experiencing reality. By means of a language system that is familiar to them, students can contribute to the expression of new points of view, experience multiplicity and contradictions, find their own words and then use them to describe their perceptions and feelings, while at the same time feeling included in the process of co-construction of knowledge in the classroom.

So, the historical excursus conducted so far shows a tension between an idea of science as a body of knowledge to be acquired and mastered, and an idea of science as a process that intersects and interacts with the lives of the students and their communities. In this regard, students are not only holders of knowledge that can be shared, but they are also actors in the complex scenario of science–society interactions. In addition, such actions are inextricably linked to an understanding and preservation of the complex ecological web that supports our life (Sachs 1999). The understanding of these interconnections requires discussion of the epistemological and methodological assumptions underpinning scientific research and education, and it involves a reflection on the responsibilities of who is in control of and manages these two crucial sectors of society (Camino et al. 2007).

In order to bring together all such aspects of ecological awareness and civic competence, scientific literacy and personal action, it is important to develop a holistic perspective that brings together different forms of societal organization and citizens' participation, with discussion of an array of approaches to sustainability. These must encompass different levels of scaling the problem, depending on disciplines (Holling et al. 1998), as well as different epistemologies, cultural values, practices and ontologies (Folke et al. 1998). In relation to responsible citizenship, we can also see that there are very different life conditions and possibilities of choice between different social groups (Gadgil and Guha 1995). Not all groups are allowed to participate equally and to be responsible at the same time. What can science education offer them, with the aim of helping them to become responsible citizens and to be involved in democratic decision processes?

In other words, educational practices can be placed at the core of a societal and cultural restructuring, which can start from the perception of scientific literacy as a more general and civic competence.

TOWARD A SUSTAINABLE EDUCATION

In the following chapters, we will illustrate some of the activities that stem from the effort to apply such concepts. We will be describing experiences that have taken place in different contexts, with a variety of educational strategies. While addressed to people working in the science area, the majority of our activities are of an inter- and transdisciplinary nature, and they all share the characteristic of bringing together scientific understanding with reflective and interactive/participatory approaches. Discussion of such approaches will focus on the development of competences, ranging from cognitive to methodological, linguistic and operational, which can be acquired as part of a sustainable education.

BIBLIOGRAPHY

Aikenhead, G. 2006. *Science education for everyday life: Evidence-based practice.* New York: Teachers College Press.

Camino, E., G. Barbiero, and A. Benessia. 2007. Abitanti globalizzati e abitanti localizzati di un pianeta messo in crisi dagli umani. Cornice teorica e piste di ricerca didattica. *Azione nonviolenta* agosto–settembre: 14–23.

Capra, F. 2002. *The hidden connections: Integrating the biological, cognitive, and social dimensions of life into a science of sustainability.* London: Doubleday.

Erduran, S., and M.P. Jiménez Aleixandre, eds. 2008. *Argumentation in science education. Perspectives from classroom-based research.* Dordrecht, The Netherlands: Springer.

Folke, C., F. Berkes, and J. Colding. 1998. Ecological practices and social mechanisms for building resilience and sustainability. In *Linking social and ecological systems,* ed. C. Folke, F. Berkes, and J. Colding, 414–36. Cambridge: Cambridge Univ. Press.

Gadgil, M., and R. Guha. 1995. *Ecology and equity.* London: Routledge.

Gallopin, G.C., S. Funtowicz, M. O'Connor, and J. Ravetz. 2001. Science for the 21st century. From social contract to the scientific core. *International Social Science Journal* 53 (168): 219–31.

Hodson, D. 1999. Going beyond cultural pluralism: Science education for socio-political action. *Science Education* 83 (6): 775–96.

Hodson, D., and R. B. Prophet. 1993. Why the science curriculum changes. Evolution or social control? In *Teaching science,* ed. R. Levinson, 22–37. London: Falmer Press.

Holling, C.S., F. Berkes, and C. Folke. 1998. Science, sustainability and resource management. In *Linking social and ecological systems,* ed. F. Berkes, C. Folke, and J. Colding, 342–62. Cambridge: Cambridge Univ. Press.

Larochelle, M., and J. Désautels. 1991. The epistemological turn in science education: the return of the actor. In *Research in physics learning: theoretical issues and empirical studies,* ed. R. Duit, F. Goldberg, and H. Niedderer, 155–75. Kiel: Institute for Science Education.

Lederman, N.G. 1992. Students' and teachers' conceptions of nature of science: A review of the research. *Journal of Research in Science Teaching* 29: 331–59.

Orr, D. 1992. *Ecological literacy*. Albany: State Univ. of New York Press.

Osborne, J. 2003. The role of argument in science education. In *Research and the quality of science education*, ed. K. Boersma, M. Goedhart, O. De Jong, and H Eijkelhof, 367–80. Dordrecht, The Netherlands: Springer.

Roth, W.-M., and J. Désautels, eds. 2002. *Science as/for socio-political action*. Counterpoints: Studies in the Postmodern Theory of Education. New York: Peter Lang.

Roth, W.-M., and A. Calabrese Barton. 2004. *Rethinking scientific literacy*. New York: Routledge Falmer.

Sachs, W. 1999. *Planet dialectics*. London: ZED Books.

Solomon, J. 1993. *Teaching science, technology and society*. Buckingham: Open Univ. Press.

———. 2004. Review of *Science education for citizenship: teaching socio-scientific issues*, by M. Ratcliffe and M. Grace (2003). *International Journal of Science Education* 26 (15): 1895–7.

5 Sustainable Education

Stephen Sterling

EDITORS' INTRODUCTORY NOTES

This chapter introduces the second part of the book dedicated to education, and it opens with a vision for change. The author points to the paradox of envisaging societal change through an education that is itself in need of rethinking. So change, evolution and transformation are the central keywords. The chapter proceeds through questions and the presentation of points of view that are different from conventional ones, casting light on the paradoxes and contradictions of particular positions (including those espoused by prestigious institutions) about what constitues a "sustainable education," which is seen as more than just 'education for sustainability'. The author introduces the notion of levels of learning, suggesting that educational actors need to engage in at least second order learning, a kind of thinking which is able to reflect on itself and so provide the basis for a shift of perspective and for change in educational practices. Within the discussion of second and third order learning we can see a link with the role played by the ideative function of language, as discussed in Chapter 4, in freeing the mind from the taken for grantedness of everyday actions.

EDUCATION IN THE CURRENT PARADIGM

> The shape of the global future rests with the reflexivity of human consciousness—the capacity to think critically about why we think what we do—and then to think and act differently. (Raskin 2008, 469)

It seems that "education" is a slow learner. In a period of unprecedented change and global threat to ecological, economic and social security, the policy and practice of Western and Westernized education are still largely built on the assumptions and epistemology of a previous age, rather than fully responsive to the conditions and needs of our time. If this is the case, then reliance on education as a critical path to a more secure and liveable future seems a risky strategy. Such a future will not be assured without learning: the question is whether formal education can and will be part of this learning. The answer hangs on whether the educational community—policy makers, theorists, researchers and practitioners—can itself experience some quality of transformative learning and awakening so that the education provision that in turn then evolves can be transformative, rather than, as at present, conformative. This chapter explores the meaning, nature and possibility of a transformative kind of education that is commensurate with, rather than lagging behind, the contemporary conditions of complexity, uncertainty, stresses and unsustainability that are increasingly familiar to all. I call this *sustainable education,* and distinguish this from "sustainability education" (which it includes) because it implies a necessary change of culture in education as a whole. This change is:

> one which develops and embodies the theory and practice of sustainability in a way which is critically aware. It is therefore a transformative paradigm which values, sustains and realises human potential *in relation to* the need to attain and sustain social, economic and ecological wellbeing, recognising that they must be part of the same dynamic.(Sterling 2001, 22)

As I have argued, sustainability:

> implies a change of fundamental epistemology in our culture and hence also in our educational thinking and practice. Seen in this light, sustainability is not just another issue to be added to an overcrowded curriculum, but a gateway to a different view of curriculum, of pedagogy, of organisational change, of policy and particularly of ethos (Sterling 2004, 50).

I arrived at this position gradually over the course of more than thirty years of working in environmental and sustainability education—increasingly understanding that sustainability logically necessitated a deep questioning and learning response in educational thinking and practice just as it does in myriad other human activities, whether economics and business, design and construction, agriculture and energy, trade and aid, health and welfare and so on. Ever since the UN Conference on the Human Environment (Stockholm 1972), education has been identified in international conferences, reports and agreements as a critical key to addressing environment and development issues. Yet, nearly four decades later—and incredibly, perhaps, given that all education is in some sense about the future—most education still makes little or no reference to these issues. And where it does, the radical implications of sustainability for educational thinking and practice are rarely addressed. In higher education—which is the sector wherein I work—an interest in sustainable development is a recent occurrence still at the margins of debate and practice, despite evidence of increasing interest and an incipient sense of urgency (Holmberg and Samuelson 2006; M'Gonigle and Starke 2006; Gough and Scott 2007; HEFCE 2008). In an international report synthesizing progress in six Western European countries, Wals suggests, "at present most of our universities are still leading the way in advancing the kind of thinking, teaching and research that . . . accelerates un-sustainability" (2008, 31). Meanwhile, a growing debate about the relationship between education and sustainability tends to be generated from within the accustomed parameters of what is considered the norm in educational policy and practice, and turn on how such norms can accommodate the concerns of sustainability. Such debate has a place and a value, but the issues that face contemporary society are of such import that we need to start elsewhere if we are to get beyond the self-referential limits of much of the debate and arrive at an adequate vision of education that can be an effective part of a more humane, participative and ecologically sustainable future. That place is the consciousness that has led to the global problematique that faces us.

BEYOND DUALISM: REVISIONING EDUCATION AND LEARNING

It has been said that the holistic medical practitioner asks a fundamentally different question than that of the conventional medical practitioner. Instead of asking: "What sort of disease does this person have?" he asks: "What sort of person has this disease?" If we apply this to the problem of unsustainability, before asking the conventional question, "What sort of malaise affects Western society?" and employing all the instrumental "fix it" problem/solution thinking that goes with it (which tends to include education), we need to first ask a deeper question: "What sort of society, or what sort of mindset, has this malaise?" So, I argue that we need to look at the nature of the crisis itself and then work backwards to examine the implications for revisioning education and learning. According to Gregory Bateson (1972)—and to many others since—our worldview is founded upon an "epistemological error," a perception of, and belief in, separateness that in turn manifests separateness and fragmentation. As Bateson (1972, 463) states:

> I believe that (the) massive aggregation of threats to man and his ecological systems arises out of errors in our habits of thought at deep and partly unconscious levels.

Bateson adds:

> Epistemological error is all right . . . up to the point at which you create around yourself a universe in which that error becomes immanent in monstrous changes of the universe that you have created and now try to live in. (461)

Global warming, to consider just one critical systemic issue, comes to mind as some kind of proof of Bateson's predictive vision. Bateson's insight stands as a radical challenge to the individualism, anthropocentrism and dualism of most Western philosophic traditions. Yet, the paradox is that—as we've become unconsciously adept at maintaining a dualistic separation of ourselves and nature, economics and ecology, subject and object, present and future—the systemic interconnections between issues have become increasingly apparent and insistent, gradually forcing, perhaps, a more holistic way of seeing the world and a questioning of long-held assumptions. Commoner's "laws of ecology" (coined around the same time that Bateson was writing) include his "First Law," which states that "everything is related to everything else" (1971, 29), and as we struggle with, say, the links between energy use, transport, climate change, global trade, food security, safety and health and the cost of living, the reality of this "law" is beginning to become more widely appreciated than when Commoner first wrote.

Bateson defined epistemology as an operative way of knowing and thinking that frames people's perception of, and interaction with, the world. Similarly, Milbrath (1994, 117) describes worldviews as "epistemological structures for interpreting reality that ground their picture of 'reality' in their own construction." The contemporary challenge to our culture is to understand itself (Clark 1989); to understand that fundamentally, the problem is not primarily "out there," but "in here," rooted in the underlying beliefs and worldview of the Western mind (Laszlo 1989). There is, then, a need to recognize the habits of thought and perception that characterize our consciousness in order to be able to move beyond them. In brief—and I have written in depth elsewhere on this (Sterling 2003)—this will involve an increasing "second-order" knowing or critical reflexivity about the dominant technoscientific, objectivist and instrumental rationality that pervades our thinking and exploring a more relational, ecological or participative consciousness appropriate to the deeply interconnected world that we have created. This entails a shift of emphasis from relationships largely based on separation, control, manipulation, individualism and excessive competition toward those based on participation, appreciation, self-organization, equity, justice, sufficiency and community.

Such ideas are given coherence in the growing literature and discourse, which suggests the emergence of a postmodern ecological worldview (Zweers 2000). This takes us beyond the root metaphor of mechanism still at the heart of the modernist mentality and toward a new organicism based on a "living systems" (Elgin 1997), "co-evolutionary" (Norgaard 1994) or "participative" (Reason and Bradbury 2001) view of the world. Such thinking goes beyond the blind alley of relativism that postmodern deconstructionism presents and toward what I would call *relationalism*—a systemic or ecological view associated with revisionary postmodernism (Griffin 1992) and supported by the implications of complexity theory (Capra 2003). Hence, I have suggested (Sterling 2007) that we might perceive ourselves as living within, and struggling to realize, an historic transition from the still-dominant modernist paradigm of realism into the idealist/constructivist position or moment, but at the same time, sensing the presence, meaning, importance and implications of an emergent, synthesizing and superseding postmodern ecological worldview or *zeitgeist*. It involves developing a collective and connective consciousness that is *holistic:* "how does this relate to that?"; "what is the larger context here?"; *critical:* "why are things this way, in whose interests?"; *appreciative:* "what's good, and what works here?"; *participative:* "who is being heard, listened to and engaged?"; *systemic:* "what are or might be the consequences of this?"; *creative:* "what innovation might be required?"; and *ethical:* "how should this relate to that?", "what is wise action?", "how can we work toward the inclusive well-being of the whole system?" Evidence of this emergent paradigm can be seen in burgeoning manifestations such as ecophilosophy, social ecology, ecopsychology and creation spirituality, as well as more practical expressions in major areas of human endeavor, such as holistic science, ecological economics, sustainable agriculture, holistic health, cradle-to-cradle production, adaptive management, ethical investment and fair trade, ecological design and architecture and efforts to develop sustainable communities.

Let us now turn to education. Applying a systems point of view, educational systems can be seen as subsystems of wider society (Figure 5.1). This being so, educational systems inevitably reflect and, to a degree reinforce, the dominant cultural paradigm.

And so, we see the continuing presence of the fundamental building blocks of the prevalent education epistemology—reductionism, objectivism, materialism, dualism and determinism—reflected from the cultural milieu and exerting an influence in purpose, policy and provision, as well as in educational discourse. These habits of thought might not be consciously recognized by most practicing educators, but they are no less powerful. They reside in the subterranean geology of education, invisible in themselves but manifested in the educational landscape above the surface: single disciplines, separate departments, abstract and bounded knowledge, belief in value-free knowing and a reluctance to deal with ethical matters in the curriculum, privileging of cognitive/intellectual knowing over affective and practical knowing, prevalence of technical rationality, transmissive pedagogy, analysis over synthesis and an emphasis on first-order or maintenance learning, which leaves basic values unexamined and unchanged. Hence, it can be argued that education shares in Scharmer's sense of a "massive institutional failure: we haven't learned to mold, bend, and transform our centuries old collective pattern of thinking, conversing, and institutionalizing to fit the realities of today" (Scharmer 2006, 3).

Three Levels of Learning

This presents those with an interest in "education for sustainability" (including global efforts under the UN Decade of Education for Sustainable Development)

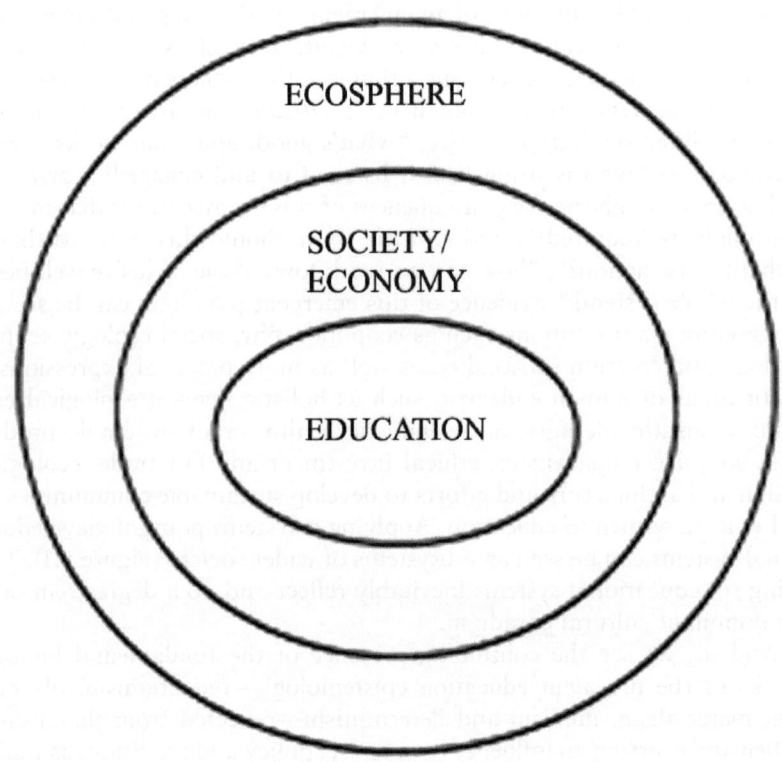

Figure 5.1   Education, society and ecosphere as nesting systems.

with a profound paradox: education is held to be a key agent of change, and yet is largely part of the unsustainability problem it needs to address. How do we work toward transformative learning in a system that itself is intended to be the prime agency of learning? This presents a double learning challenge. It is helpful to elaborate the distinction made between these two arenas of learning; that is, between "structured learning"—the designed learning associated with courses and programs for students—and the social or organizational learning within institutions that needs to take place in order to facilitate the former (Sterling 2006): deep change in the former depending on deep change in the latter. Here, we can draw insight from learning and organizational change theory, and I have discussed elsewhere the usefulness of staged change models, again partly based on Bateson, distinguishing between first-order, second-order and third-order learning (Sterling 2003, 2007). First-order learning is essentially a noncritical, adaptive, response (to the concerns of sustainability in this case) based on the values and *modus operandi* of instrumental rationality. Second-order change, by contrast, is:

change that is so fundamental that the system itself is changed. In order to achieve (this) it is necessary to step outside the usual frame of reference and take a meta-perspective. (Ison and Russell 2000, 229)

While third-order or epistemic change is desirable but difficult, I would contend that individuals, educational institutions and whole-educational systems need to at least have some experience of second-order change or higher-order learning—a questioning and revisioning of basic assumptions in order to facilitate a sufficient response to the urgent challenges of sustainability.

AN ECOLOGICAL FRAMEWORK FOR SUSTAINABLE EDUCATION

To help encourage this process of rethinking and revisioning, I have endeavored to suggest and map out the framework of a sustainable education paradigm that tries to respond to, embody and articulate an ecological worldview in its educational thinking and practice. I see this as offering the possibility of a unifying theory of education and learning that integrates the best of past liberal education practice with newer emphases, such as transformative learning, capacity building, creativity and adaptive management, considered part of the new sustainability agenda. The rest of this chapter summarizes some of the key points of this work. Using the term "sustainable education" was quite deliberate: I wanted people to move from "how do we educate for sustainability or sustainable development?" toward deep attention to *education* itself—its paradigms, policies, purposes, and practices, and its *adequacy* for the age we find ourselves in. In other words, I wanted the focus of attention to move from a response typically around *provision* (which often amounts to a partial tweaking of the curriculum), to *paradigm*, the epistemic sets of values and ideas that fundamentally influence curriculum thinking and design and all the other aspects of educational provision. In fact, I make a distinction between four "Ps", which can be seen in systemic interrelationship, as shown in Figure 5.2.

These Ps are relevant to any systemic level from the national system to the institutional level—and even the departmental level within the institution—a disciplinary set, for example, forming a subsystem of an institutional set. Using this model, it is possible to outline possible "whole system" shifts from the conventional system to a more ecologically based one.

An important point needs to be made here. Gough and Scott (2007, 16) interpret this paradigmatic argument by suggesting that "in this view . . . education's role is to promote the new paradigm." This misses the point. Rather, I argue that educational thinking and practice needs to evolve to *become*, embody and, at the same time, critically elaborate this paradigm. The challenge here is whole-system transformation—not "promotion of a view"—which needs to take place in a co-evolutionary way in tandem with changes in wider society. Thus, the traditional and

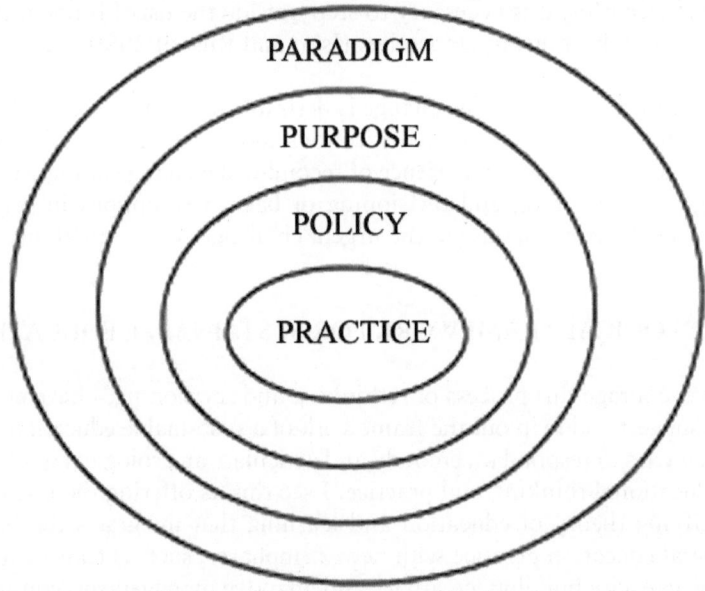

Figure 5.2 The nesting four "Ps" of education systems.

Paradigm Instead of education reflecting a paradigm founded on a mechanistic root metaphor and embracing reductionism, positivism
and objectivism, *it begins*: to reflect a paradigm founded on a
living systems or ecological metaphor and view of the world,
embracing holism, systemisism and critical subjectivity. This
gives rise to a change of ethos and *purpose . . .*

Purpose Instead of education being mostly or only as preparation for
economic life, *it becomes*: a broader education for a sustainable society/communities, sustainable economy and sustainable ecology. This expanded sense of purpose gives rise to a
shift in *policy . . .*

Policy Instead of education being viewed solely in terms of product
(courses/materials/qualifications/educated people) *it becomes*:
seen much more as a process of developing potential and
capacity through life at individual and community levels
through continuous learning. This requires a change in methodology and *practice . . .*

Practice Instead of education being largely confined to instruction
and transmission, *it becomes:* much more a participative,
dynamic, active learning process based more on generating
knowledge and meaning in context, and on real-world/situ-
ated problem solving.

(Based on Sterling 2004)

simplistic question of "how can education change people's behavior?" that has informed much education for change in the past is superseded by a systemic view, which generates a very different question: "how can education and society change together in a *mutually affirming* way, towards more *sustainable patterns* for both?" (Banathy 1991, 129). This signals a change from education focusing on maintaining the existing state and operating as a rather closed system towards helping shape society "through co-evolutionary interactions, as a future-creating, innovative and open system" (129).

This is echoed in the idea of "critical engagement" between the university and wider society. Fear et al. (2006) suggest that such engagement is:

> an enduring, ever-unfolding, and enfolding process of experiential learning . . . (it) is persistent, critical, reflexive, discursive, inclusive, pluralist and democratic (179).

I used the word "sustainable" in the term "sustainable education" to imply four descriptors that apply to educational values, policy and practice: "sustaining," "tenable," "healthy" and "durable."

- *Sustaining*—in process and outcomes, it helps sustain people, communities and ecosystems.
- *Tenable*—it is ethically defensible, working with integrity, justice, respect and inclusiveness.
- *Healthy*—it is itself a viable system, embodying and nurturing healthy relationships and emergence at different system levels.
- *Durable*—it works well enough in practice to be able to keep doing it, through continuous critical reflexivity and renewal.

It is perhaps a useful exercise to evaluate current systems and specific institutions against these simple but demanding indicators. To elaborate further, sustainable education:

- "implies a fundamental change of purpose, or at very least, an additional key purpose for education
- implies embedding, embodying and exploring the nature of sustainability as intrinsic to the learning process. This is education "as" sustainability—nurturing critical, systemic and reflective thinking, creativity, self-organisation and adaptive management—rather than simply education "about" sustainability, or education 'for' particular sustainable development outcomes
- is not prescriptive, but is indicative and purposeful
- affirms liberal humanist traditions in education, but goes beyond them through synergy with systemic and sustainability core values, concepts and methodologies

- challenges the limiting effects of characteristics of the dominant mechanistic paradigm such as top-down control, centralisation, managerialism, instrumentalism and the devaluing of humanities and arts
- is based on "systemics" rather than "systematics", that is, the emphasis is on *systemic learning* as change, rather than *systematic control* in response to change." (Sterling 2003, 278)

Mapping out some of the features of this view gives us more detail:

Box 5.1   Some implications of a systemic view of education and learning.

A shift from:

- fixed knowledge *toward* recognizing uncertainty and "other ways of knowing"
- decontextualized and abstract knowledge *toward* applied and local knowledge
- emphasis on cognitive experience *toward* valuing affective, inspirational, intuitive and practical knowing
- valuing intellect *toward* also valuing intuition
- information and data *toward* deeper knowledge and wisdom
- curriculum control *toward* curriculum subsidiarity and negotiation
- teaching *toward* learning
- content *toward* process
- restricted learning styles *toward* multiple learning styles
- passive instruction *toward* participative and critical enquiry
- uncritical learning *toward* reflexive learning
- selection and exclusion *toward* social inclusion
- formal education *toward* learning for life
- specialists *toward* generalists in teachers and learners
- individualism *toward* organizational, community and social learning
- institutional isolation *toward* social and community engagement
- single and separate disciplines *toward* more inter- and transdisciplinarity
- instrumental values *toward* a new integrative sense of social/ecological ethics and responsibility
- competitive values *toward* cooperative values
- placelessness *toward* celebration of place
- valuing "knowing" *toward* valuing "being"

(Based on Sterling 2003, 271)

The choice here is not between binary opposites, but a suggestion that sustainable education involves a change of weighting away from the dominant paradigm in a way that transforms and conserves some of its characteristics rather than jettisoning them in their entirety—even if that were possible. What is needed is a more adequate, encompassing paradigm, but one that can evolve from where things stand now rather than present an unattainable ideal.

As I have argued, this paradigm needs to be an ecological or relational one because of the systemic nature of the world we inhabit. A sufficient and whole-learning response to sustainability is required at three levels—personal, organizational and social—and in the three interrelated areas of human knowing and experience. These are *perception* (or the affective dimension), *conception* (or the cognitive dimension) and *practice* (or the intentional dimension). In each of these areas, higher-order learning toward an ecological consciousness and competence involves movements toward:

"respons-ibility" *an expanded and ethical sense of concern/engagement;*
"co-rrespondence" *a closer knowledge match with the real world; and*
"respons-ability" *the ability to take integrative and wise action in context.*

This holistic, three-part model of paradigm change can be understood in various ways. In academic language, it can be seen in terms of *epistemology, ontology* and *methodology*. In traditional educational terms, it represents the *heart, head* and *hands* of the individual learner. In everyday terms, it underlines *awareness, understanding* and *competence,* and it points to the learner who is at once *concerned, connected* and *capable*. In strategic terms, it represents developing *vision, critique* and *design* for change. In educational terms, it represents *paradigm and purpose, policy and curriculum* and *pedagogy and practice*.

It should be clear at this point that the idea of sustainable education is a larger, deeper and more reconstructive notion than "education for sustainability" or "education for sustainable development." While I acknowledge and welcome the much higher profile that "education for sustainable development" (ESD) has achieved in recent years—bolstered significantly by the UN Decade of ESD—this apparent progress throws into relief an uncomfortable tension between accommodatory and radical transformist approaches. There remains confusion in the higher education sector about the qualitative difference between "embedding sustainable development in education"—most often an accommodatory "add-on" response that does not necessarily impinge on or challenge existing norms—on the one hand, and a reformative or transformative change as more holistic response involving cultural change and whole institutional shifts on the other (Sterling and Scott 2008). There is a parallel here with sustainable development discourse in wider society: either it is an add-on area of theory and practice—that is, a sectoral interest, which can be safely considered *alongside* but separate from other agendas—OR, to a greater or lesser degree, it involves all aspects of social, economic and political organization. So it is with education in the service of sustainability. As I have argued previously, sustainable education shifts attention from "adding on" some desired learning outcomes as in ESD (although this has its place not least as a first step) toward thinking about the kinds of education through which sustainability qualities and well-being manifest as emergent properties in the institutional and wider communities.

In suggesting this, I am arguing that sustainability is not primarily something that we need to educate *for* in the preparative sense, but rather it is a matter of embodying a healthy relational approach to teaching, learning, organization and community, and this takes us to a much closer identification between the notions of sustainability and *educere*. Sustainability is:

> to do with appreciating and respecting what is already there, with both conserving and developing inherent creative potential, with assisting self-reliance, self-realisation, self-sustaining abililties and resilience. From this perspective, it is not difficult to see the parallels between, or the integrative pattern that connects, ecologically sustainable development practice and sustainable education—that connects "becoming more sustainable" and "becoming more human" . . . Such principles as diversity, relative autonomy, community, and integrity have an echo in both natural and human contexts. It is only a very short jump to see how educational values such as differentiation, empowerment, self-worth, critical thinking, cooperation, creativity and participation are part of this picture. (Sterling 2001, 35)

Given current socioeconomic and ecological conditions, and a widespread sense that our patterns of living will change either by choice or circumstance, such thinking seems much more acceptable, necessary and practicable than it did a few short years ago. Indeed, outside formal education, some are positively embracing a vision of "energy descent" and a localized and greener future, and the idea of transition communities manifesting these alternative futures is spreading (Hopkins 2008). Formal education has much catching up to do to match such innovation and this will inevitably involve some real challenges. As Fear *et al.* (2006) suggest:

> We need to get beyond the notion that issues of pressing concern to the citizenry are social, *or* economic, *or* moral, *or* political. Instead each is a slice of overarching and interconnected complexities . . . of contemporary life . . . (and) of the same messy reality. These matters are not mere semantics, but essential and systemic challenges to the dominant worldviews and prevailing worldviews of the Academy, as well as to the very paradigmatic nature of the academic activities that reflect them. (180)

CONCLUSION

The Great Transition Initiative, a global think tank quoted at the start of this paper, suggests that "the momentum toward an unsustainable future can be reversed but only with great difficulty . . . (and yet) . . . a planetary transition toward a humane, just and ecological future is possible" (Raskin et al. 2002, 95). If education is not part of the critical reflection and path-finding this inevitably involves, one might reasonably wonder what it is for in our current volatile age. This chapter has attempted to suggest elements of a map that might help in the task of reorienting educational thinking and practice toward serving the purpose of assuring a safe, liveable and sustainable future. The articulation of an ecological view of education and learning hopefully contributes to critical discourse and, in some way, acts as a vision or attractor that can make its realization more likely.

BIBLIOGRAPHY

Banathy, B. 1991. *Systems design of education.* NJ: Educational Technology Publications.

Bateson, G. 1972. *Steps to an ecology of mind.* San Franscisco: Chandler.

Capra, F. 2003. *The hidden connections.* London: Flamingo.

Clark, M. 1989. *Ariadne's thread—The search for new ways of thinking.* Basingstoke, UK: Macmillan.

Commoner, B. 1971. *The closing circle.* New York: Bantam Books.

Elgin, D. 1997. *Global consciousness change: Indicators of an emerging paradigm.* CA: Millenium Project.

Fear, F. C. Rosaen. R. Bawden, and P. Foster-Fishman. 2006. *Coming to critical engagement.* Lanham, MD: Univ. Press of America.

Gough, S., and W.A.H. Scott. 2007. *Higher education and sustainable development: Paradox and possibility.* London: Routledge.

Griffin, D. R. 1992. Introduction to SUNY series in constructive postmodern thought. In *Ecological literacy: Education and the transition to a postmodern world,* e. D. Orr. Albany: SUNY Press.

HEFCE. 2008. *Sustainable development in higher education—Consultation on 2008 update to strategic statement and action plan* London: HEFCE.

Holmberg, J., and B.E. Samuelsson, ed. 2006. Drivers and barriers for implementing sustainable development in higher education. Technical paper no. 3 Education for Sustainable Development in Action. Göteborg Workshop, December 7–9, 2005, 61–7. Paris: UNESCO

Hopkins, R. 2008. *The transition handbook—From oil dependency to local resilience.* Totnes: Green Books.

Ison, R., and D. Russell. 2000. *Agricultural extension and rural development—Breaking out of traditions, a second-order systems perspective.* Cambridge: Cambridge Univ. Press.

Laszlo, E. 1989. *The inner limits of mankind.* London: One World.

M'Gonigle, M., and J. Starke. 2006. *Planet U. Sustaining the world, reinventing the university.* Gabriola Island, BC: New Society Publishers.

Milbrath, L. 1994. Stumbling blocks to a sustainable society. In McKenzie-Mohr, D. and M. Marien eds. Special issue, "Visions of Sustainability," *Futures* 26(2):

Norgaard, R. 1994. *Development betrayed—The end of progress and a co-evolutionary revisioning of the future.* London: Routledge.

Raskin, P., T. Banuri, G. Gallopin, P. Gutman, A. Hammond, R. Kates, and R. Swart. 2002. *Great transition: The promise and lure of the times ahead.* Boston: Stockholm Environment Institute/Tellus Institute.

Raskin, P. 2008. World lines: A framework for exploring global pathways. *Ecological Economics* 65: 461–70.

Reason, P., and Bradbury, H., ed. 2001. *Handbook of action research—Participative practice and enquiry.* London: Sage Publications.

Scharmer, O. 2006. Theory U: Leading from the future as it emerges—The social technology of presencing. *Fieldnotes* September–October, Halifax, NS: The Shambhala Institute for Authentic Leadership.

Sterling, S. 2001. *Sustainable education—Re-visioning learning and change.* Schumacher Briefing no. 6. Dartington: Schumacher Society/Green Books.

———. 2003. *Whole systems thinking as a basis for paradigm change in education: Explorations in the context of sustainability.* PhD diss., Centre for Research in Education and the Environment, Univ. of Bath. Available online at www.bath.ac.uk/cree/sterling.htm

———. 2004. Higher education, sustainability and the role of systemic learning. In *Higher education and the challenge of sustainability: Contestation, critique, practice, and promise,* ed. P.B. Corcoran and A.E.J Wals. Kluwer Academic.

———. 2006. Towards sustainability intelligence. Paper presented at the seminar Developing innovation competences: What does Waginengen University (want to) offer?, Waginengen, November 26, 2006. Also available online at http://www.tad.wur.nl/uk/newsagenda/archive/news/2006/Seminars.htm

———. 2007. Riding the storm: towards a connective cultural consciousness. In *Social learning toward a more sustainable world: Principles, perspectives, and praxis,* ed. A.E.J. Wals Waginengen, The Netherlands: Waginengen Academic Publishers.

Sterling, S., and W. Scott. 2008. Higher education and ESD in England: A critical commentary on recent initiatives. *Environmental Education Research* 14 (4): 386–98.

Wals, A., ed. 2008. *From cosmetic reform to meaningful integration: Implementing education for sustainable development in higher education institutes—The state of affairs in six European countries.* Amsterdam: DHO.

Zweers, W. 2000. *Participation with nature—Outline for an ecologisation of our worldview.* Utrecht, The Netherlands: International Books.

6 Science Education for Sustainability
Teaching Learning Processes with Science Researchers and Trainee Teachers

Elena Camino, Giuseppe Barbiero and Daniela Marchetti

In practice, institutions of education exist for teaching, rather than for learning or for learning to learn. What is learnt, above all, is information, routines and obedience, in other worlds, facts, procedures, and to do what one is told. [. . .]

(Chambers 1997, 62)

EDITORS' INTRODUCTORY NOTES

This chapter presents some educational experiences conducted with postgraduate students engaged to become teachers and professional scientists. Such experiences can be seen as the practical application of the epistemological and educational reflections presented in Chapters 1 and 4, and they have the specific aim of supporting young people in the development of the required competences for building a participatory democracy and a sustainable society.

With respect to the experiences with future researchers, particular attention will be given to the relational aspects—in the interaction among lecturers and between lecturers and students—which have been quite interesting for us, but very little is written about this in the current literature. In relation to the courses with future science teachers, attention is focused on describing in some detail some of the activities that are proposed in the teacher education courses, with a view to illustrating the endeavor of building coherent links between learning and teaching processes and "transformative" educational aims, as described in Chapter 5. It was also our concern to bring to light and emphasize the variety and breadth of content—both disciplinary and transdisciplinary—that it is possible to deal with through this kind of reflexive and interactive educational approach.

TOWARDS AN EDUCATING COMMUNITY

What choices of content and methodologies can teachers make when they want to address the issues raised by the new global era—both in science and society? This chapter looks at some of the educational approaches that our Science Education Research Group has introduced and tested over the years, with the aim of providing young people with the competences for becoming citizens of the "expert democracy" that was illustrated in Chapter 2. Our efforts were directed toward a transformation from a reductionist and objectifying view of the natural world toward a holistic and integrated view (Bateson 1973; Capra 2002; Gallopin et al. 2001; Manghi 2004; Sterling 2001, 2002 and Chapter 5). This view stems from an idea of the subject inquiring into the world, considering its aspects of ignorance, boundary and dependency, and valuing the sense of inclusion, wonder and respect—a view that is also in line with a post-normal science epistemology (Funtowicz 2001; Funtowicz and Ravetz 1999; Gallopin and Vessuri 2006).

The main principles informing our worldview and pedagogical practice are briefly summarized here:

a. Humanity is included in the natural systems and is totally dependent on them.
b. Natural systems have evolved over a long period of time and their structures and relationships carry the traits of their evolutionary path. This is currently expressed through their biodiversity, an ongoing result of both continuity and innovation of living processes interacting with the abiotic environment.
c. The relationships that humanity establishes with the environment are expressed through a multiplicity of channels, from the cognitive to the emotional. These, in turn, are shaped by the culture and the socioenvironmental context, so that what we know and how we know about natural systems is continuously shaped according to our worldviews, which nourish powerful narratives (Benessia, Chapter 1).
d. Because the Earth system is limited and interconnected, every reflection and every social practice can only be deemed democratic if it takes into account the link between ecology and equity.
e. Awareness of the consequences of one's actions is a necessary element for a sustainable life.

In the planning and implementation of educational activities, we gradually developed the approach of incorporating different elements and blending them in different ways—the idea of a bricolage (Chambers 1997; Kincheloe and Berry 2004), depending on participants' age, contexts and expectations. We can briefly summarize these as follows:

• A "basket" of choices: the underlying idea is that each person with whom we establish a relationship has different attitudes, interests and also things that one might not appreciate. All these aspects need to be taken into account and respected. If our educational offering is rich and varied, then it is more likely that each person will find something motivating that can help him or her to grow and develop confidence in his or her own abilities.

- A variety of strategies: silent, personal reflection, which allows one to get in contact with his or her deeper self; dialog and open exchanges with others to appreciate the variety of approaches and interpretations, and to reorganize one's own conceptual maps; working in small groups to cooperate toward a shared aim; searching for links between disciplinary knowledge and everyday events in order to learn to contextualize scientific knowledge and make sense of it in everyday life (Aikenhead 2006).
- A multiplicity of relational approaches within the group, which are useful in transforming hierarchical relationships into relationships of equivalence (Patfoort 2006), and may support the introduction of a nonviolent context (Galtung 1996). For such a transformation to occur, great care is taken in clarifying that, while people have different roles and responsibilities, it is important to respect and value the variety of characters, ways of expression and interests of each person.

In particular, our group has devised and widely tested role-plays on controversial socioenvironmental problems (Colucci Gray et al. 2006 and Chapter 8). These are complex activities, rich in opportunities with regard to knowledge and competences, as well as the multiplicity of relational approaches that are offered. In addition, such simulations not only deal with complex and controversial socioenvironmental problems and lead students to investigate implicit epistemic assumptions, but they also allow the tackling of the crucial aspect of conflict, proposing experiences of nonviolent transformation that are necessary in a framework of sustainability.

THE EDUCATION OF FUTURE RESEARCHERS

A few years ago, we were given the opportunity by the Piedmont Region of Italy to design and implement a course on sustainability education addressed to young doctoral students and researchers from the University and Polytechnic of Turin, who were involved in research on environmental themes. The Region had funded the researchers' studentships, and for this reason, they were interested in supporting our proposal of promoting, through an innovative course, a deeper knowledge of sustainability topics, and setting the basis for building a team of young scientists who were able to enter into dialog with one another and undertake interdisciplinary research.

The course ran for eighty hours, twenty of which were devoted to a residential stage in the mountains. It was an experience of great interest for all, students and teachers. Details of the organization of the course have been published elsewhere (Camino et al. 2005), and the final evaluation report is available from the Interdisciplinary Research Institute on Sustainability (IRIS) website[1]. In this chapter, we do not elaborate on the content of the course, neither do we focus on the students' learning process. Rather, we spend some time reflecting on the work of planning and coordination that involved a few of us as tutors in the course.

The majority of teachers/tutors were members of IRIS. The Centre has among its objectives not only the exchange of information, but also the promotion of dialog among different disciplines as a premise for the development of interdisciplinary research. While some of us had already entered some form of collaboration—i.e., Bravo, on the concept of Gaia as global commons (2004), and Bagliani et al., on a critical approach to the Kuznet curves (2008)—for other people, it was a new experience. Planning a course together did not imply reaching a shared idea of sustainability, but it certainly called for making explicit the research methodologies and the set of models, strategies and viewpoints—in other words, the interpretive schemes of each discipline—as well as their implicit assumptions and paradigms. Moving from a simple exchange of information (i.e., the results of empirical research or the references related to a topic) to a more intense dialog involving criteria for choices, discussion of the efficacy and reliability of methods and the inferential processes that were used to compare variables and produce results, generated controversies and many situations of tension. The efforts to listen to one another with humility and the desire to understand other researchers' perspectives were alternated with expressions of unease and mutual attitudes of de-legitimization and dismissal. It was interesting, if sometimes difficult to accept, to become aware that the so-called scientific premises of each discipline are carriers of worldviews and value-laden choices that were deeply rooted. Often, these assumptions would make it impossible to overcome controversies and resolve conflict.

In trying to understand environmental controversies, it does not make much sense to look for "what science really says". "Even the most apparently apolitical, disinterested scientist may, by virtue of disciplinary orientation, view the world in a way that is more amenable to some value systems than others. That is, disciplinary perspective itself can be viewed as a sort of conflict of interest that can never be evaded" (Sarewicz 2004, 392).

All the tutors involved have benefited from a great learning experience. The participation of a language specialist (M. Dodman, one of the authors of Chapter 4), a Jungian psychoanalyst and some scholars involved in artistic performances allowed the group of tutors to widen their perspective. The exchanges and interactions among them promoted a deeper insight into their own as well the others' disciplines, from both a methodological and an epistemological point of view. In addition, there was an opportunity to pose new questions to oneself about the processes of construction of new knowledge; about the role played by ignorance and the critical and crucial relationships between science, society and law within the perspective of a sustainable future (as reported in Chapters 1 to 4).

With respect to the relationship with the students who had been involved in an interactive and reflective educational relationship, we report here some comments from the tutors:

> Caretto & Spagna (Artists). We needed to meet with the students and the other members of the IRIS group, to bring a deeper critical reflexion on our personal artistic research and the relation between it and the theme of sustainability, and the possibility of contributing to a fruitful discussion within the group. We had to clarify with other members of IRIS the role that we artists are trying to play as regards the links between art, science and sustainability issues. Not only at a conceptual level, but also in practice. [. . .] Students were keen to dialog with us about the controversial links between art, science and sustainability.

> Perazzone and Tonon (Natural Scientists). [. . .] our two lessons did not aim at the construction of new knowledge by the doctoral students (as some of them probably knew more than us about the carbon cycle!). Our idea was that of casting light on the complexity of the real in relation to the inevitable limitedness of our interpretive schemes and our modes of representation . . .

> Giunti (Regional Park Ranger). I tried to pass on the idea that the understanding of such complexity [of the ecological systems] is a fundamental premise—at least at the level of attitudes and dispositions—for making territorial and political decisions both for the long and the short term and scale.

Students were offered a broad overview of the meanings of sustainability (Tukker 2008), and of the approaches that the different disciplines have elaborated in order to deal with sustainability issues. Moreover, students were given the opportunity of developing metacognitive competences (Bateson 1973; Varela et al. 1991) for dealing with problems by going beyond the empirical evidence, and above all, for accepting—at least in part—the sense of limitedness of human knowledge and the importance of creativity and intuition in the production of new knowledge.

Doctoral students appreciated the fact that some tutors were walking the path along with them, witnessing that they were also engaging in a process of learning: doubt, critical self-awareness and acknowledgment of error were valued and practiced by all participants. At the same time, it was useful to refer to the synthesis made from time to time by the more experienced tutors. This helped to make links between themes and perspectives that were, in appearance, very different from one another.

Some doctoral students remarked that a total change of mindset had occurred during the course, and their view of things had changed. At the beginning, it was not clear to them that the course was aimed at providing

Politics

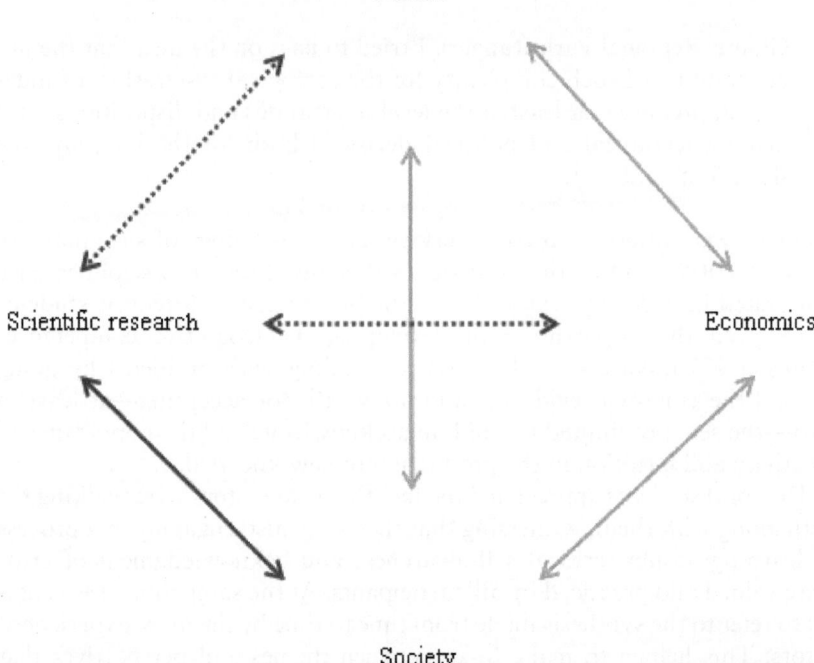

Figure 6.1   Two examples of "synthesis." (a) a cartoon that allowed participants to recall a debate on controversial issues. (b) a map underlying links between different perspectives.

new conceptual tools and promoting new interpretive schemes rather than simply transferring ideas. This had been said and repeated several times but it was not easy to understand. Only through experience and participation did they begin to make sense of it and appreciate it.

- At the beginning, it seemed difficult to link the information we were receiving with our own research field; only gradually it became "automatic".
- Numerous cues for reflection were offered to integrate the different aspects: a mental operation which I had not considered before and which is now occurring spontaneously.
- At the beginning I was asking myself what was the point of being there; I had difficulties in linking the topics with my specific discipline (chemistry). But then, when I understood that the barriers could be taken down and that there were significant links between science and art (which I used to keep strictly separate), I became very interested and I wanted to explore deeper.
- Listening to concepts which are used with different meanings by different tutors helps to de-familiarize some words, so that they are not too rigidly connected to "objects" or processes.

THE EDUCATION OF FUTURE TEACHERS

The Inter-University School of Specialization (SIS)

The members of the Science Education Research Group and IRIS lead several courses as part of the teacher training school in natural sciences. This school is characterized by two main educational strands: preparing teachers of mathematics, chemistry, physics and general science for the lower secondary school, and teachers of natural sciences, chemistry and geography for the upper secondary school. Students enrolled in a teaching degree at the SIS hold a science degree (i.e., natural sciences, chemistry, physics, biology, geology, mathematics etc). Generally, biology and natural sciences postgraduates, enrolling for the teaching diploma in natural sciences at the upper secondary level, have a sufficient knowledge base for school teaching. In contrast, Italian school organization assigns only one person to teach both math and scientific disciplines (physics, chemistry, biology, geology) to students 11 to 14 years of age (lower secondary school). So, mathematics graduates, as well as any science graduate from the fields listed previously, can access this teaching job. As it is clearly an impossible task to adequately teach sufficient math to natural scientists or teach science to mathematicians within the two-year span of the SIS school, the science courses we are responsible for are perceived as an opportunity for introducing windows into transdisciplinary dialog. Through this approach, each participant shares his or her knowledge with a view to comparing and eventually integrating the different disciplinary perspectives.

Action-Research and Participation

The transdisciplinary research, as well as the teaching and learning activity carried out by our research group is characterized by the practice of Action-Research (Marchetti 2008; McNiff 2002): "Transdisciplinary research is a form of Action Reseach. Participation and learning cycles have to start from the beginning" (Haberli et al. 2001, 9).

We are in a situation in which the researcher is internal to the process: the teacher-researcher is part of the same teaching and learning process that is being observed, and he or she is often supported by an observer who is responsible for recording the events, the dialogs and the activities that take place.

Data consist of observations, written down by the teacher-researcher or the observers during the plenary sessions or the group activities; products from the students, elaborated during the interactive and reflective activities introduced during the course; individual responses to questionnaires, group worksheets and end-of-course assignments. Such data is also a form of feedback that impacts the course itself; for example, with regard to the planning of future courses and, at the same time, with a view toward building theoretical models of more general validity. The evaluation of the impact of the formative offering on the students is conducted at the end of a process of data triangulation, involving the observer, the teacher and the students themselves. The students in particular are always asked to comment on what had been proposed and how they felt and participated as a form of self-evaluation. Through such means the relationship with the participants becomes more inclusive and equal—as is documented in some of the comments from the students.

This approach also offers the possibility to promote involvement from the participants and to develop an active relationship between the researcher and the "objects" of study. Within the framework of Action-Research, the "reflexive capacity" is a central element: this is intended both as a form of self-awareness on the side of the researchers (awareness of one's own preconceptions, implicit curriculum, expectations and educational aims), and as an attitude and competence to be promoted in the participants/students.

Research is therefore conducted in "real world" conditions; in other words, open systems in which people, information and situations are continuously evolving, just as happens in every learning and teaching process. Because of the educational context in which we operate, research has both an "exploratory" and "emancipatory" nature: the objective is that of trying to reach a certain level of understanding of what happens as a result of the stimuli that had been introduced, yet without losing sight of the primary intention, which is that of offering students the opportunity to develop new competences and abilities. Research is guided at the same time by an idea of education as an evolutionary and shared process, and by specific problems to be solved or to be overcome through a collective effort.

Unveiling Approaches

For a few years, our research group has been reflecting on the possibility of helping future researchers and teachers tackle the fragmentation and compartmentalization of disciplinary knowledge. The reorganization and integration of such knowledge by the learner would make for a more coherent acquisition of new knowledge, and also a more efficient use of such knowledge for understanding oneself, natural processes and phenomena and socio-environmental problems (Marchetti 2008).

One of the strategies that has proved useful in "making order" in the broad and often messy field of the learners' knowledge has been that of identifying and bringing to the learner's attention the different approaches underpinning the disciplines in the natural sciences. As reported previously in this chapter, scientists—when looking to a new problem—are always guided by previous knowledge and assumptions of their specific discipline, and they (often tacitly) apply ways of looking phenomena and processes that affect the choice of significant variables and the methods for data-gathering and elaboration. From a student's point of view, becoming aware of what are the interpretive schemes guiding scholars in exploring the problems that are typical of their own discipline can be extremely useful for understanding the motives that guided, as well as for making the objectives of the inquiry explicit.

All experimental sciences (but most explicitly natural sciences) are characterized by a variety of different approaches (Mayr 1982):

- the *descriptive* approach looks at structures, allowing for comparison and classification of life forms on the basis of their similarities.
- the *functional* approach investigates the dynamic processes that allow the harmonic balance between change and invariance.
- the *historical* approach is set up for reconstructing the possible scenario of a unique story, which is both individual and collective, and which unfolds along various time scales (e.g., from human to geological).
- the *systemic* approach (Odum 1997) is concerned with revealing the web of connections that sustains each life form (including human life), and which connects them within a hierarchically organized whole.

Each academic discipline (e.g., anatomy, physiology, ecology etc.) emerges from applying a prevalent approach to natural systems. Scientists are aware that each approach is necessarily both limited and limiting, and each one is framed within historical, cultural and linguistic contexts (Wittgenstein 1969; Longino 2002). Any scientific undertaking makes use of the specific epistemological status of each discipline, which is applied as a flexible and changeable tool (Dodman 1999). Moreover, specialized languages have been created and are continuously transformed in order to support each particular view (Chapter 4).

Clarifying the chosen perspective is also a way of making apparent the limits of any approach: attention to structural details requires cutting out the part that is being studied from its relationships or interactions with the other parts. Conversely, such relationships are crucial for those who are interested in emphasizing the mutual interactions occurring between the different elements of the systems. Some investigations require the simultaneous measuring of events occurring within a short time, but obviously, synchronic measures need to be integrated by diachronic measures in order to reconstruct the time course of single processes.

So, the different "ways of seeing" (descriptive, functional etc.) allow scientists to identify, frame and separate, either physically or conceptually, objects and processes from their context. This procedure is based on the definition of boundaries, which are, in different cases, either understood as barriers, preventing exchanges between contiguous compartments, or as privileged surfaces, intersected by flows of matter and energy. Equally, practicing with synchronic and diachronic readings (Arcà and Guidoni 1987), comparing different time and space scales (Gallopin et al. 2001) and looking for patterns (Capra 1997) can help us to develop "conceptual tools" for a transdisciplinary approach to the observation and interpretation of the living systems. As we will illustrate in the following sections, a dynamic and fluid application of conceptual tools can be—in our opinion—an effective way for revealing the nature of each single approach (Volk 1998; Maturana and Varela, 1998), and for integrating the various approaches with one another.

Using Conceptual Tools

Awareness of the variety of approaches and conceptual tools used by scientists may help students to realize that science progresses not only by selecting the objects of study, but also the conceptual categories through which the world is interpreted (Cini 1994). This process can help them to produce a less naive and more mature view of the nature of science (Aikenhead and Ryan 1992).

Recognizing the plurality of perspectives and the limitations of each one is also helpful in developing awareness of the systems that we are studying, and in which we are at the same time embedded as an intimately interconnected part. A reorganization of knowledge made on the basis of the various approaches can help reordering, but it may not be sufficient for recognizing the complement of each approach and promoting integration between them. In this regard, we have experienced the use of particular concepts that proved useful for this particular purpose. Such concepts can be used as "tools" supporting the development of mental connections, bridges between the different perspectives and approaches. These are not new ideas: quite simply, we try to identify and implement methods that are easy to understand and put into practice, which can support the construction of a framework in which different systems of thoughts (and the accompanying knowledge) can enter into dialog with one another.

In the following sections, we provide examples of two conceptual tools: the idea of "boundary," and the way of looking at the natural systems through the lenses of "energy flows and matter transformations." On a number of occasions, during courses addressed to future secondary school teachers, we have devised and tested activities through which we aimed at developing competences in the use of such mental instruments. These activities usually require a couple of hours and they take place according to the methodology described earlier: brief moments of individual introspection alternated with work in small groups, plenary discussions, deepening of concepts, metareflections. Usually, such activities begin with an open question and they end with a multiplicity of answers and new questions. Indeed, the objective of these activities is not only that of providing a more complete and articulated overview of a scientific concept or issue, but also developing abilities in finding new connections and posing questions in an autonomous and creative fashion. On several occasions during such courses, discussion and reflection are focused on the strategies, the choice of content and the objectives and evaluation criteria of the educational activities that the students will, in turn, propose to their own pupils. For reasons of space, we will not discuss this particular aspect in this text.

The Concept of Boundary

The concept of boundary is profoundly rooted in each one of us: it originates from the experience of perception, and it is part of the primary metaphors of our cognition (Lakoff and Johnson 1999). The continuous shuttling between the literal and metaphorical meaning makes it a very powerful conceptual tool, which is also easy to use by children and young people.

Awareness of our disposition toward seeing and setting up boundaries, and at the same time, the double meaning of the concept (boundary as a way to separate parts or to allow exchanges), makes this concept a useful tool for connecting different approaches.

Boundaries arise from the choices of the researcher: to trace boundaries is never a neutral act. The choice of boundaries isolates elements and processes we want to study, and cuts out links and flows that connect "internal" and "external" parts of the system. Such a choice depends on properties known or attributed to the object under test.

A clear example is that of the cell. The concept of "cell" was born following the identification of the boundaries that surround some "cells", but it was the difficulty in delineating such structures that led to various discussions among biologists in the nineteenth and twentieth centuries (Virchow 1856; Mazzarello 1999). When studied by the morphologist, the cell membrane separates the internal environment from the external one: this compartmentalization allows electrochemical gradients to be maintained, which are essential to life processes. But it is also true that the cell membrane is the crucial point of passage and exchange between inside and outside: because of the sophisticated characteristics of selectivity of channels, pumps and carriers, life functions can take place, such as the transmission of electric signals or fast ionic flows.

Once applied to the cell, the concept of boundary can become a conceptual tool to explore other borders at different scales: for example, what are the boundaries of an ecosystem? This question has been asked during courses with future teachers as a trigger for reflection (Camino et al. 2002), and it always generates a sense of puzzlement. The majority of participants provide a structural definition (the limits of the woods, the margins of the beach around the pond[2]): only a few people are able to put forward an answer that is in line with the relevant discipline—ecology—which focuses on relationships and flows. Starting from the definitions offered by the student teachers, it is possible to generate an interesting discussion, reasoning around the nature of hierarchical systems, asking about the difference between ecosystems and biomes, looking at disciplinary perspectives and approaches. The discussion can also be enriched by short explorations of textbooks, readings of scientific articles and interviews.

> Among ecologists willing to draw *any* lines between ecosystems, no two are likely to draw the same ones. Even if two agree, they would recognize the inherent artificiality of their effort, and probably make the attempt with only a few species in mind. [. . .] Different lines are not surprising, but rather are entirely expected, because of the intrinsic interconnectedness of living systems (Corn 1993): *the discrepancies between scientists accurately reflect the diversity of the real world* (author's emphasis).

> Ecosystem processes are scale dependent and, as such, *the choice of boundaries for an ecosystem is of profound importance to the conceptualization of an ecosystem* (author's emphasis) and the scope and validity of questions being asked within that ecosystem. The process–function approach [. . .] addresses the functional role of constituent parts of ecosystems and, therefore, is often organized around understanding the cyclic causal pathways that maintain ecosystem functions. Energy flow and biogeochemistry are points of focus for ecosystem ecology under this approach (Post et al. 2007, 112).

Once the concept of boundary has become a conceptual tool, it can be applied to other fields and it can be unveiled where it is not explicitly defined. We can cite two recent cases in which the ambiguity in the definition of boundaries generated scientific controversies of great relevance given the socioeconomical implications. The first case refers to the production of biofuels, which has divided the scientific community: on one side, there are those who think it is an effective solution for reducing greenhouse effects; on the other side, there are those who maintain their inefficacy. Nobody had done their calculations wrong, but each side reasoned according to different boundaries (Patzek and Pimentel 2005).

Reflecting on boundaries helps future teachers to not only reason about ontological aspects (the interconnected and interdependent nature of real world) but also on epistemological aspects:

> A useful practice in scientific research would be to always define the system within which we isolate or delineate the problem investigated, and to look for relevant interlinkages. In other words, look outwards to examine how the issue/problem is linked to other variables, issues or systems (horizontal and vertical or cross-scale linkages), in time and space. Only then we can meaningfully ignore the rest of the system (if the linkages are negligible) or decide how, and to what degree, to include the broader system in the research. (Gallopin et al. 2001, 228)

A fluent competence in applying the concept of boundary can also help us move more easily along changes of scale. For example, what are the boundaries of our body and what do they consist of? Dealing with the topic of nutrition by looking at all the barriers that the molecules encounter in their journey through the organism helps to understand many processes: from the boundary of the mouth, perceived as the entry door, and where, through ingestion, food disappears from our perception to the long, slow and laborious process of breakdown and digestion. This process reduces food particles to the acceptable dimensions for getting through the "real" boundary—the intestinal walls—with their extraordinary selective properties.

Physically different but conceptually similar is the barrier that the molecules of air, oxygen, carbon dioxide and nitrogen have to go through, as they are also dealing with cell structures. Hence, the idea of boundary allows us to connect not only the knowledge developed from the different approaches (in the examples above we looked at structural, functional and systemic approaches), but it can also facilitate the conceptual leap from the explanations of processes taking place at different scales.

In conclusion, reflection on boundaries takes us back to a fundamental strategy for knowing: it depends on us wanting to see the world as all interconnected or divided. After all, if it is true that everything is connected, as some maintain, in order to make sense of our "dialog" with the world, it is important to be able to organize our view of the "undifferentiated phenomenic flux" (Cini 1990) by identifying and labeling parcels bounded in space and time; in structures or levels, phenomena, events or processes. In so doing, we can try to understand how some things are connected to others and to define, even if arbitrarily, at least something about the nature of their relationships (Arcà 1993, 1995).

Energy Flows and Matter Transformations

In a time when energy is one of the most talked about topics in society, it is sad to see how little schools and universities try and help young people to understand more about this concept. Particularly with respect to answering some key questions (Where does it come from? Why is it "consumed"? Why do we have an increasing need of it?) and, above all, to learn to link energy problems with other issues, which are closely connected, such as the transformation of natural systems, water consumption, the role played by food habits and international commercial trade.

School and university education in Italy provides a very fragmented, disciplinary approach to the study of energy. This often means that young postgraduates attending teaching qualification courses do not have the opportunity for a dialog among them about energy, even if they share a science background (if they are physicists, chemists, geologists or biologists). Each one of them refers to definitions, interpretations, physical variables and models that are so different that it seems they are speaking about processes and phenomena that do not have anything in common. There are interesting studies and educational reflections on this problem, which, unfortunately, are unknown to disciplinary specialists: in particular Keith Ross (2000a, 200b) and Doménech et al. (2007).

Smil (2003, 2008) presents the theme of energy through a historical perspective, and connects theoretical aspects with quantitative data and socioeconomical reflections. What is particularly interesting is Smil's effort to link energy use and ethical issues through a rigorous and quantitative approach to lifestyles:

> rich evidence leads to the conclusion that the average consumption of between 50–70 GJ/capita provides enough commercial energy to secure general satisfaction of essential physical needs in combination with fairly widespread opportunities for intellectual advancement and with respect for individual freedoms (2008, 352).

These data, accompanied and supported by a huge variety of information, lead to the writing of a striking conclusion to the book: "shaping the future energy use in the affluent world is primarily a moral issue, not a technical or economic matter. So is the narrowing of the intolerable quality of life gap between the rich and the poor world" (Smil 2008, 370).

Thus, a serious and detailed, scientific quantitative analysis leads us to confirm concepts that had been previously expressed in qualitative form about a century earlier by Gandhi: "Earth provides enough to satisfy every man's need but not every man's greed."

Given the scientific and social relevance of energy issues, it seemed important to us that our educational research included a study of the topic of energy. Among the various activities that we have tried out, some have proved particularly effective: for example, the presentation and shared reflection on the "scale of energies," which allows us to appreciate the enormous range of orders of magnitude through which energy can manifest itself, as well as the equally large variety of physical descriptors that have been introduced for the purpose of defining and measuring it. Through a reflexive conversation with the student teachers about the scale of energies, it is possible to find points of contact between the approaches of physicists and chemists (who think in terms of electron volts), physicists and biologists (with the conversions between calories and joules) or botanists and ecologists, who are dealing with the comparisons between the energy that is captured during each photosynthetic process and the energy that is gathered by the entire terrestrial biomass (Volk 2001): "the key concept is embodied energy, the portion of solar energy that comes to reside in the bodies of photosynthetizers as chemical energy and that is used to fuel the metabolism of other organisms" (157). The extraordinary dance of energy and matter, and the mutual influences and interactions connecting life and abiotic environment, are of crucial importance for understanding not only how the natural systems work, but also ourselves within such systems, as we have managed to increase our power from the modest 50–60 watts of our basic metabolism to 10,000 or more watts (Smil 2003), thanks to the availability of numerous "energy slaves." In the United States, each person has the equivalent of 100 energy slaves working 24 hours a day for him or for her.

Thinking together with future teachers about flows of energy and matter transformations—first through simple and well-defined cases, and then extending the reflection and exploring time and space scales of higher and lower orders—allows us to acquire a powerful conceptual tool that we can use to reorganize our ideas, makes us aware of our actions and inform our choices.

In the following sections, we will describe some examples of activities carried out in the context of one of the courses addressed to future teachers of mathematics and science in the lower secondary school. The activities that are described here are relatively simple: once more, they can be enriched and made more complex depending on the way they are being proposed and on the level and degree of students' participation—individual reflexion, exchanges of points of view, dialog, formulation of new questions, updating through recent scientific research, social and economical implications . . .

Behind the History of Things

The reflection that inspired the activity we describe here stems from two considerations. On the one hand, many science teachers still hold tight to a transmissive teaching style. They also continue to use explanatory language, by means of which natural processes and phenomena are described as simple, noncontroversial facts. This deprives students of the opportunity of being genuinely involved and expressing their points of view as individuals contributing to knowledge construction. On the other hand, the central role played by narration in learning processes has been widely described by many authors (e.g., Cladinin 2006), and it is considered again in this book in Chapter 8.

In our courses, we have tried to give space to narration: we started from the observation that often, the stories of "objects" are focused on their movements in space and time, from the past to today, as if they were products with stable identities. We wondered if, by means of different modes of narration, we could shift attention from products to processes (from conservation to transformation, both at the micro- and macroscopic level) and from matter to energy (i.e., to the causes of movements and transformations).

Some familiar objects are taken to class and randomly distributed among the participants (some are natural objects; others are artificial ones); for example: a bird's feather, a piece of plastic, a pine cone, the plastic wrapping of a mail packet, a bottle full of mineral water, a seed. In the first part of the activity, students are invited to work individually and write about the following points (each request is given only after the previous task has been concluded):

1. Tell the story of . . .
2. Extend from past to future
3. Tell a new story in terms of processes of transformation
4. Identify the causes of such transformation

This activity is welcomed with curiosity by participants, and it is usually carried out with interest. By comparing the different stories, it is possible to appreciate the variety of ways in which the students have interpreted their task, the tendency toward looking at the products or the processes, their creativity. The written texts are great stimuli for connecting different perspectives: on the one hand, there is the story of the dead leaf "which lived in a park, it fell from the tree when it was not yet completely grown, and it was carried by the wind to the feet of a person who picked it from the ground"; on the other hand, another contribution refers to transformations of a plastic spoon with an industrial past and connections to the oil factories, the oil rigs and even further past, the sedimentation and diagenetic origin from a very ancient plant . . .

Through the search of what is behind the stories, it is possible to make links with the transformations of matter and energy that were necessary to produce them. These are the premises for acquiring the concept of Life Cycle Assessment, a technique to assess the energetic and environmental aspects associated with a product, process or service by compiling an inventory of relevant energy and material inputs and environmental discharges, and evaluating the potential environmental impacts associated with identified inputs and outputs.

The sharing of the texts written by student teachers, and the discussion that follows, helps to cast light on the role of energy (in its various forms) in any process of matter transformation, either in the living world or the world of manufactured objects. Knowing such concepts can help inform choices in everyday life. For example, it is sufficient to think about the increasing number of products that are now sold with the logo "ecolabel". The close interactions between energy and matter also emerge, and together it is possible to touch upon the concept of eMergy: this is the energy required directly and indirectly to make something, or the energy of one type that is embodied in any form of energy, good or service (Odum 1998). Up to now, we are not aware of such a concept being introduced in university courses and known to our students, even those holding a degree in physics. Yet, it seems crucial nowadays to reason not only on quantitative aspects, but also about the quality of energy (which was the motive behind Odum introducing this new concept).

Reasoning about eMergy offers the opportunity to analyze significant themes in ecology, which are often dealt with in a superficial and inappropriate manner at school. According to most textbooks, the food chain and the biomass pyramid can be explained as a linear energy transformation chain; at each step, some energy is degraded and some is passed to the next step. These concepts can be clarified and enriched through the emergetic perspective that reveals qualitative as well as quantitative aspects. A teacher aware of the implications of such knowledge will be able to develop among his, her young students an appropriate reflection on this fundamental topic in ecology, which is too often reduced to a mechanical scheme in school textbooks. Energy flows along a food chain are interconnected with huge matter use and transformations, as with carbon dioxide production and water consumption[3].

The City Under the Dome

The conceptual tool of energy flows and matter transformations can help connecting processes that take place on very different scales of time and space. Often the explanations of textbooks keep such processes strictly separate because they are dealt with by different disciplines. Aspects of thermodynamics and ecology can be integrated with the physiological perspective through the concept of metabolism. Here, we briefly describe the main points of an activity that was taken from the book by Wackernagel and Rees (1996): the book that signaled to the general public the birth of the now widespread concept of the ecological footprint (EF).

Participants are invited to form small interdisciplinary groups, then they are asked to answer some questions:

1. What would happen to any modern city [. . .] if it were enclosed in a glass or plastic hemisphere that let in light but prevented material things of any kind from entering or leaving?
2. Then, let's assume that such a city is surrounded by a diverse landscape in which cropland and pasture, forests and watershed are represented in proportion to their actual abundance on the Earth, and that adequate fossil energy is available to support current levels of consumption using prevailing technology. Let's assume our imaginary glass enclosure is elastically expandable: How large would the hemisphere have to become before the city at its center could sustain itself indefinitely on the land and water ecosystems and the energy resources contained within the capsule? (Wackernagel and Rees 1996, 9–10).

The teaching/learning context and approach are as already described, alternating reflection with small group work, debates and collective discussions. By starting from the exercise described here, we have built up a learning experience that allows participants to become aware of the extraordinary increase of energy flows, transformations and transfers of matter that have become possible by means of technoscientific development and the expansion of industrial societies.

The move from the almost closed ecosystem (in terms of matter cycles) of ancient human settlements—from the prehistoric to medieval ones (King and Monger 1986)—to the open ecosystems of modern cities has implied a progressive increase in the EF, alongside a parallel impoverishment of the availability of goods and services for "peripheral" populations.

This activity allows future teachers to reflect on two important and complementary aspects: on the one hand, the total dependency of humanity on global natural systems (the inhabitants of a city "under a dome" would perish within a few days); on the other hand, the relationships between ecology and equity. In a closed system, such as the Earth, with a fairly constant flow of energy from the sun and a limited availability of matter and natural processes, the perspective of equal distribution of resources for satisfying everyone's needs is critical.

Real Leaf, Fake Leaf

The capacity for artificially reproducing natural "objects" is extraordinary: artificial flowers can be produced that are almost indistinguishable from the real ones. It is not unusual when entering the lobby of a hotel to be welcomed by a wonderful fichus tree . . . made of plastic! The eye, perhaps, is satisfied, but something subtle has changed.

Student teachers are invited to make a list of the similarities and differences between a real leaf and a fake leaf, which is identical with regard to shape, color, dimensions, thickness and consistency, but it is made of plastic. This activity leads future teachers to apply the interpretive approaches of their respective disciplines (physics, chemistry, natural sciences) in order to carry out the task. As often happens when proposing activities that appear quite simple in the beginning, this exercise progressively generates a series of open questions and mutual learning. This happens because the language and models of physics (i.e., the selective absorption of electromagnetic waves by molecular structures) cannot be easily integrated with the perspective of chemistry (i.e., the absorption function of chlorophyll: "120-plus atoms arranged into a binary structure [. . .] absorber extraordinaire used by virtually all photosynthetizers" [Volk 1998, 128]). Even among the same group of biologists, there are those who choose a structural approach and those who prefer looking at the functional aspects about the occurring processes. The different disciplines guide the choice of different space scales: there are those who look at the relationships between the molecules, yet neglect the macroscopic morphological and functional aspects (the stomata, the lymph), and the role that the macrostructures play in the processes of photosynthesis and respiration: the study of such processes makes quite apparent the inseparability of living beings from their context and their mutual modifications. Active debates arise about the fluctuation of temperature in the real leaf and in the fake leaf. The approach of thermodynamics (the vibrational levels of molecules) is compared to the interpretive schemes of biology (the homeostatic systems of control, water movements etc.). As the analysis develops, the differences between the characteristics emerge, as well as the stories and the functional properties of the living tissue of the leaf as compared to the thin plastic sheet.

Thanks to the dialogic approach of the educational process, the compartmentalized vision of the graduates is enriched and made fluid. Identifying the flows of energy and matter transformations at the different space and time scales acts as a powerful conceptual tool that helps to grasp the complexity and the interconnections between the different levels of the natural systems in which both we and the objects of our attention are a part. This competence is very much a necessary one for the citizens of the global world. It helps them to grasp connections that often are not made explicit. For example, it drives us to grasp the relationship between energy production and water consumption[4], or between the power supplied by an engine and the amount of exhaust gas, or even between food choices and energy input.

Global Issues

The conceptual tool of energy flows and matter transformations can be applied to a great variety of processes, phenomena, situations and systems. It is not a matter of looking exclusively for "science" topics; on the contrary, everyday life, both at the personal and collective levels, can be effectively explored through this conceptual tool. We refer here to two activities. Both were undertaken by starting from two open questions expressed respectively in a graphic and iconic form. The activities were carried out as previously described: an initial moment of personal reflexion, then sharing, discussion and gathering of further information until new connections are made and new questions are formulated.

Oil Eaters

> Some authors maintain that we—inhabitants of modern industrial societies—can be defined as "oil eaters." Why? In your opinion, is the sentence to be interpreted literally or figuratively?

This activity is part of a research strand that we have been pursuing for many years and that is looking at the level of awareness that science teachers have of the role that the natural sciences can play in promoting understanding of socioenvironmental problems. Thinking in terms of energy flows and matter transformations in following the chain of processes of food production and consumption can be very useful for understanding that the consequences of the energy crisis are not only manifested in the transport and industry sectors. A reduced availability of oil can have a dramatic impact on global food production. In vast areas of the planet, in fact, this is totally dependent on fossil fuels to provide petrol for machinery and harvesting systems, and it is also dependent on their byproducts for the production of fertilizers and pesticides. "The most damaging, dangerous and certainly the least noticed aspect of the contemporary food system is the extent to which the supply of even the most basic food has become dependent on petroleum" (Jones 2001).

Indeed, future science teachers to whom we have proposed these activities have so far shown to be generally unaware of the dependency of the modern agricultural system on nonsolar energy inputs. By reflecting with them on the energy flows and matter transformations in the processes connected with food production, it gradually emerges that it takes energy not only to transform matter, but also to acquire, transport, store and even use energy. Such invested energy may be compared to "returned energy," and a new powerful conceptual tool can be applied: EROI (energy return on investment); that is, the ratio of the energy delivered by a process to the energy used directly and indirectly in that process (Cutler 2004).

This concept can be simplified, made usable also by younger pupils and applied to illuminate some inconsistencies of our affluent society:

> E_o/E_i expresses the ratio between the energy content of food product and the energy that was required for producing, processing, packaging and preserving it. By simple calculations, we can discover that in traditional and pre-industrial societies, E_o/E_i is approximately equal to 100; for the products of mass distribution, E_o/E_i can shrink to values that are even less than 1! (Jones 2001).

Reflecting on the relationships between food and oil can be used to widen the discussion to include the extraordinary possibilities acquired by techno-science for the transformation of matter by means of an increasing power density (W/m^2) and energy intensity (J/g) (Smil 2008): corn can be used in baking tortillas, as well as for ethanol in cars and power plants; natural gas can be made into fertilizers for food output. New avenues have been opened for the deployment of matter and energy, yet with some unexpected outcomes for those who did not take into account some fundamental concepts: each new usage is located within a closed system—the Earth. In such a system, the availability of matter is limited and the eXergy—that is, the capacity of energy to produce mechanical work—progressively decreases.

Interlinked Ecosystems

One of the IRIS members, Massimo Battaglia, architect and cartoonist, endeavored on various occasions to represent by means of vignettes some of the themes dealt with during our courses and stages. Some of these vignettes appeared to be particularly effective in generating open questions and therefore we used them for educational purposes. We present here one such vignette, which proved useful in relating two conceptual tools: the "boundary" and "the energy flows and matter cycles."

The cartoon was presented to all participants at the beginning of a lesson and it was accompanied by a particular task: "to give a title, write a caption

Figure 6.2   Vignette for "boundary" and "energy flow and matter cycle."

and list some topics of the life sciences which have relevance for the depicted scene". Such an iconic suggestion elicited a variety of interpretations that were provided by the participants. The analysis of the answers given to the questions from the cartoon showed a rich variety of explanations/interpretations, which provided cues on the underlying views and value systems of the participants (e.g., "natural world against modern world"; "equilibrium between production and consumption"; "North and South"). With regard to this vignette, interpretations that were opposed to one another were also given. It was interesting to see how this made an impact on the participants and made them more interested in listening to the voices of others. This was not to be taken as a premise for counteropposition and argumentation, but as the start of a growing awareness of the limitations of any single interpretation and the potential of a plurality of ways of seeing[5] (Volk 1998; Ravetz 2005; Chapters 1 and 2). Exploiting the possibility to represent metaphors and paradoxes by means of images, the vignette allowed teachers and students (after the sharing of the different perspectives) to cast attention on two elements. The first was explicit and it referred to the possibility to draw as adjacent two environments that, in reality, are geographically distant. The other (that was only subtly hinted at and was not grasped by everybody) referred to the energy flow and the transport/transformation of matter between North and South on the planet. With the extraordinary increase in international transport, both in numerical and power terms, enormous quantities of goods are transported everyday from one side of the world to the other. In this process, the closed cycles of natural ecosystems are made linear and, by opening the boundaries, they contribute to creating a unique, global ecosystem. The final, collective reflection allowed the group to develop a greater awareness of the interlinking of scientific knowledge, technological applications, energy resource use and everyday choices in modifying ecosystem boundaries (Odum 1997), as well as in redesigning the web of life at a global level. Moreover, some students enriched their final report by creating new cartoons that they proposed as examples of effective teaching tools for secondary school students.

Playing with Language

> Understanding is fluid, flowing like water,
> while knowledge is like blocks of ice that prevent the flow.
> Such is the difference between knowledge and understanding.
> (Thich Nath Hahn)

One of the risks of a transmissive approach to teaching is that of contributing to "fix" concepts in a process whereby the signifier and the signified become "stuck" together, leading to the "thingifying vision" of science (Chapter 4). The risk is greater the more the concepts concern phenomena, processes and events that escape our direct perceptual experience. Offering opportunities to reflect on concepts—even on a few, selected examples—can help students develop, not only a more adequate scientific knowledge of those concepts, but also a more general awareness that will enable them on other occasions to distinguish object from concept and understand how every concept has a historical development, is dynamic, transitory, as well as often being a vehicle for multiple meanings. Above all, it helps to recognize how concepts are powerful and flexible tools that facilitate new mental connections (*cum–capio* = I connect), not, as often happens, a rigid product to memorize.

Towards Making Concepts Fluid

The Concept of Gene

The activity begins by encouraging student teachers to write down their own ideas about the concept of gene. The ideas normally range from, most commonly, that of an object (a piece of DNA, a cluster of molecules, a particle, a structure) to that of a process (a factor that permits the expression of characteristics, a unit of genetic expression, a hereditary unit). Comparing the ideas and categorizing them during a plenary session creates the basis for further steps. First, particular linguistic features can be identified (e.g., the presence of metaphors); then, a search is performed with books, websites, recent publications or books on the history of science (e.g., Keller 2000). Each time new cues can emerge and different aspects can be investigated, depending on the interests and competences of participants. The teacher helps to develop the emerging ideas by asking open questions, underlying ambiguities, offering citations from authors embracing different views etc.

> It may well turn out that all we can say about genes is that they are continuous or discontinuous DNA segments whose precise structures and specific functions are determined by the dynamics of the surrounding epigenetic network and may change with changing circumstances (Capra 2002, 177).

Alternatively, the history of the evolution of the meaning of gene can be reconstructed, providing illustrations of the ways in which the gene has been interpreted by different authors over the past 100 years, a situation that continues today (Fox-Keller 2000, 31):

- The "gene" is nothing but *a very applicable little word*, easily combined with others, and hence it may be useful as an expression for the "unit factors," "elements" or "allelomorphs" in the gametes.
- There is no consensus opinion among geneticists as to what the genes are—*whether they are real or purely fictitious.*

- Watson and Crick convinced biologists that genes are real molecules, and this was followed by *the identification of DNA as the genetic material.*
- We are far from the idea of a self-contained, stable DNA. *DNA alone cannot even copy itself*; besides, without a complicated system of monitoring, revisioning and correction, replication would encounter many errors.
- The stability of gene structure thus appears not as a starting point, but as an *end-product*—as a *result of a highly orchestrated dynamic process* requiring the participation of a large number of enzymes organized into complex metabolic networks [. . .].

Finally the participants are asked to reformulate the concept of gene within an interdisciplinary perspective through the construction of concept maps (Novak and Gowin 1984). Here is how some trainee teachers commented on the activity:

> To our surprise we realised that our first ideas of the map were a disaster! The word gene didn't even appear and the central element involuntarily became DNA [. . .] We tried to rethink our mental schemata. Time and space are the basis for all the concepts introduced which, even if retaining some important structural and functional aspects, open up to other epistemological fields that lead to the reconstruction of the history of life, ethics, relationships with culture, medicine . . . (participant to a SIS Course, A060, 2006/07)

In this way, significant results are reached through the deconstruction and reorganization of scientific knowledge within the group, through the development of linguistic competence and the ability for epistemological reflection, and through the opportunities for applying interdisciplinary approaches.

A complementary exercise that can be proposed refers to the idea of gene as investigated through the conceptual tool of "boundary." In the science community, three models of genes are held up (Barbiero et al. 2006) and they differ on the basis of the boundaries that are established to define the gene: (1) exon model; (2) genic DNA model and (3) integrated model. According to the first one, the only significant elements of a genome are the exons—the DNA sequences transcribing for a specific gene (Crick 1958). This model does not take into account the links between gene products nor the networks with the organism: a boundary is established by translating the DNA sequence (exon) to a gene product without considering networks between gene products. The "genic DNA model" includes DNA sequences transcribed to RNA but not translated into gene products: such a model, though still excluding intergenic DNA (over 70% of the human genome), is extremely complex, and so far it has been impossible to exploit knowledge coming from this approach. Finally, the "integrated model" considers the genome as a whole, including intergenic DNA, its evolutionary process, its links with the host cell and its exact copies in all other cells of the organism. This model shifts the attention from objects (genes, proteins) to the circular relationship between organism and genome.

A Plurality of Signifiers and Meanings

Many of the words in scientific language are used to express a literal meaning, whereas the same words used in everyday language express a figurative meaning. For example, in scientific language, "vital" refers to something living, whereas in everyday language it is used to refer to something very important. There are numerous examples of this phenomenon in science education literature, and teachers should be very aware of the need to define meanings in context. With postgraduate student teachers of science, we have based a rethinking of the concept of evolution of living beings on a comparison of the different uses of the word "evolution" in everyday and scientific contexts, such as the evolution of the embryo, the psychology of early age (childhood and youth) and evolutionary biology. In one of our courses we distributed individual questionnaires: all the questions elicited some reflection upon words concerning the evolution of natural systems. We report here two questions that stimulated answers that gave rise to lively discussion and helped—through a linguistic approach—to develop knowledge and awareness of various evolutionary biology issues.

> *The words "evolution" and "revolution" are both connected to the idea of change, but express different types of change. Write two sentences using these words and explain how they exemplify the difference.*

Through a participative reflection focused on language, it was possible to enlighten some implicit assumptions held by student teachers. Most of them held a deeply rooted idea of evolution as a slow and gradual process, and were not aware of the model of punctuated equilibria, proposed by Eldredge and Gould in 1972.

> *What is meant by "species A is more evolved than species B"? Is this expression scientifically correct?*

We collected many varied answers that were shared and gave rise to much debate. The majority believed the expression correct, but with different supporting arguments: if A is more complex, more ancient, modified more times as regards ancestors, more similar to man, more specialized. Those who believed the expression incorrect affirmed that it is not possible to measure a level of evolution, or that it makes no sense to compare the evolution of different organisms.

These activities share a number of aspects: they stir up cognitive as well as emotional involvement, offer opportunities for peer education and promote genuine motivation for deepening one's knowledge of issues. After such inter- and transdisciplinary activities, bringing linguistic and epistemological reflection together with rethinking scientific concepts, the majority of the participants declared they had built new and significant knowledge and developed greater awareness of the dynamic nature of scientific knowledge itself, as well as being more motivated and ready to work on the theme of evolution with secondary school students (Cerruti 2007).

Nominal Language Hides Subjects

As we have seen in Chapter 4, nominal language has a great synoptic capacity, gained, however, at the cost of hiding processes and agents (Dodman et al. 2008). The following is a simple activity that can help become aware of implicit assumptions about worldviews (sometimes also antithetical, depending on the participants). First, there is a brief discussion of some features of language: for example, how words are sometimes used literally but more often figuratively, or how context is all-important in determining meaning. Then, in groups, students are invited to write sentences on strips of paper containing words derived from "sustain." The sentences are grouped and categorized on the basis of any criteria the students wish. Next, the different criteria are illustrated and discussed. Among the most common are usually those of literal/figurative meanings, grammatical categories and fields of use (psychological, environmental, political etc.). Finally, the participants are each asked to write a way of expressing in verbal, everyday language the synoptic expression "environmental sustainability," making agents, processes and objects explicit. Normally, two types of sentences are produced: one of which considers that human beings sustain nature, while the other states the opposite. From these differences emerge questions for reflection and discussion, with a subsequent decision to further investigate the issue from a scientific and philosophical point of view. The following are three sentences written by young scientific researchers working on environmental issues:

- The earth sustains our use of its resources, even our exploitation of them.
- We sustain each other reciprocally. The earth sustains us and we should sustain the whole system.
- Man sustains the environment by conserving the resources for future generations.

Reconnecting Outside and Inside

We have come to the boundaries of our planet—the exploration shifts from the external world to the internal world.

> Human beings live in a world which is
> in some way mysterious;
> new things that happen and which can be experienced in it
> cannot be explained and
> not only those things which happen in the realm of what is expected.
> The unexpected and the absurd
> belong to this world.
> Only then, life is complete.
>
> C. G. Jung

While it is still widely practiced, teaching based on representation and transmission, explanation and demonstration of the scientific basis of the single disciplines has limited effectiveness and negative consequences, although often involuntarily so. Transmission-based teaching is one among many strategies; it serves precise and limited purposes: for example, to introduce a new topic, summarize or reconnect to previously met ideas. A vast literature produced by research in science education has illuminated the risks of rote learning and superficial understanding. The abilities to generalize and use information acquired in other contexts are limited, and students elaborate, often unconsciously, an idea of science that is objective, neutral and that describes reality "as it is."

As a consequence, young people develop the tendency to perceive scientific knowledge as the knowledge of *something,* rather than knowledge that is socially constructed and negotiated. Teaching strategies that are heavily based on explanation and demonstration contribute to offer a *"thingifying"* view of science (Larochelle and Désautels 1991; Désautels and Larochelle 1998), which is often accompanied by a sense of alienation, if not fear, toward nature (Chapter 4).

Giving value to personal experience, the development of critical and reflexive attitudes, the openness toward listening to others, the bringing together of specialized approaches within a transdisciplinary perspective, the importance of context, the seeking of a plurality of points of view, the acceptance of limits, the awareness of the possibility of going wrong . . . in all these aspects we recognize precious elements that every teacher can use to propose educational experiences within a perspective that we can define as "post-normal" (Chapter 1).

Recognizing and appreciating the different approaches that sciences make use of to explore the natural systems, as well as using conceptual tools for integrating knowledge, can give more meaning to our vision, but this is not sufficient. These are mental operations that engage us mainly at the cognitive level, while other dimensions (which are unsaid and unrecognized) are left in the background: emotions, artistic intuition and experience. In order to help future teachers also make use of these other approaches to knowing, it is important to get them personally involved and explore the dimension of the internal self. So far, we have described steps toward the visioning and promotion of an "educating community," in which each participant feels at ease, shares experiences with peers, is willing to explore the natural systems and understand their functions. In achieving such an aim, the emotional dimension of our knowledge of nature plays an important role that, perhaps, has been underestimated.

Retrieving Memories of Childhood

We mention here briefly an activity that we have been proposing for many years to future teachers of secondary schools, with outcomes that move and encourage us. As for the other activities described in this chapter, this is a fairly simple one; however, proposing it within an academic context, giving it as much time and respectful attention as for the more traditional activities, acts as a stimulus for the participants. In particular, they are invited to reflect not only on the specific task given to them, but also more widely to ask themselves why in schools—and particularly during the hours of science—there is very little opportunity to express our most profound feelings toward nature, to talk about oneself and the emotions generated in our encounters with nature.

The task is the following: after a short moment of silent concentration, we ask student teachers to write about a vivid memory from childhood that is connected to nature and to explain why it has remained so strongly impressed in their memory. We transcribe here some of the comments:

> The only but precious memories of my childhood are the summers I spent on the Alps with my grandmother. I remember in every detail the days with the animals, the food I was eating, the games I was playing and my stick. Memories of Turin—almost none.
>
> When I was playing football in the wheat fields near my house. The wheat had just been harvested and the stingy bits were left (it was sore running over them).
>
> Afternoons spent at my uncle and aunt's country house in Sicily. A swing made of a wooden board and hanging from a tree—the wild asparaguses, the places where I was running.
>
> The color of the bluebottles which I have never seen any more in the fields. I was going looking for them on my bicycle.
>
> When I was playing with my brother in amongst the tall grass: we would dig out a kind of hole in the grass. We would stay there I don't know for how long. It was springtime, with the sun, the ants, the bees

buzzing . . . it was very nice, we would play with the grass, we would not get bored at all . . .

This activity can be reconnected to the vast literature on the role and importance of the experiences of nature in childhood (e.g., Nabhan and Trimble 1994; Sobel 1993; Thomson et al. 1994; see also Chapter 7). This is a fascinating and important theme, which the scientific education of future teachers of secondary school often does not take into account. And yet, these are experiences of crucial importance that contribute to the construction of that worldview that will shape the choices, values and even ways of doing science when becoming adults.

The now extensive collection of such memories has allowed us to identify some regular patterns. In addition, the comments that are expressed in the conversations following the phase of sharing have induced us toward drawing a few conclusions.

After an initial moment of embarrassment and wonder, almost everybody is willing to write. The memories are generally associated with complex experiences, an element of intense sensory perception (colors, smells), a human presence (children, friends and grandparents) and a dimension of doing (running, building, hiding, rolling). Such memories trigger strong emotions, a sense of astonishment for having temporarily forgotten about them and a desire to narrate them and share.

Following this activity, future teachers appear to have acquired greater awareness of the importance that such experiences have had on their lives and on developing one's ecological identity (Thomashow 1996).

In addition, becoming aware that an increasing number of children will not be able to live such moments—which, at one time, were usual and frequent for all—saddens and worries them. It is almost as if only now they are gaining consciousness of the gravity of the loss caused by the urbanization processes in children's psychophysical development.

Scientific knowledge is interwoven with worldviews; it is shaped by them and informs them in turn. In the path toward sustainability, it is therefore important to develop both vision and knowledge together.

> To act well, we need to experience the Earth not as "nature" out there, nor as an "environment" that is distinct from us, but a mysterious extension of our very own sensing bodies that nourishes us with an astounding variety of intellectual and aesthetic experiences (Harding 2006, 244).

The Voice of the Protagonists

In a perspective—as we have repeatedly underlined—that proposes to the students, future teachers the opportunity to participate in an active and reflective manner in the learning and teaching process, their words are perhaps the most appropriate means for concluding this chapter:

- The interdisciplinary and reflexive approach proposed during the course is in my opinion useful for putting into perspective our position as human beings: nature is not dominated by us, neither is benign or malignant towards us. This is not because nature is an entity which is indifferent to us, in a Universe which is even more indifferent. Rather, it is because nature is not an entity which is separate from us, or better, we are not separate from her, but we are part of a single system.

- Perhaps our peculiarity is the fact that we can become conscious of ourselves. In this case the word we can use is mandatory: in truth, we are aware of being able to impact on the system, we are somewhat conscious of how much we are effectively acting upon it but totally unconscious of how much the system is actually impacting upon us.

- I liked the phrase which came out during one of the first lessons that the "world is non-disciplinary", that is it does not belong to any discipline and therefore each discipline are glasses which can be used to observe the world and see different things depending on the lenses (of the chemist, the physicist or the biologist) which are used. It is fascinating to know that the same event can be studied by a mathematician and by a biologist because one point of view does not exclude the other, rather, they can complement each other.

- A striking aspect is in my opinion that of the uniqueness of living beings: what characterizes each life form—a human being, a single flower, even an insect—is indeed the fact of being exclusive and impossible to repeat. [. . .] A lesson in the Life Sciences can be a suitable context for initiating a reflexion on emotionally involving matters. In this regard I would find it appropriate to tackle the theme of the diversification of the living, introducing the concept of biodiversity. In addition, this could be an opportunity for talking about the importance of being different, which is not a limit, but a noteworthy opportunity. This can make us reflect also on the differences that exists within the human species, the differences of personality, capabilities, interests which should not isolate people but bring them together, with a view of cooperation and sharing of resources.

- Twenty days after I had experienced the birth of a new life (my first baby), participating in the course of didactics of the life sciences has been an emotionally charged experience. In those months I have been experiencing and I still do such an intense psychophysical change that my reaction to such lessons has been totally unexpected. As a student

I would find the study of biology quite boring. I had received a traditional education, based almost exclusively on a systemic-descriptive approach. In the course of such lessons I have realized that biology is a complex, dynamic and changing science, which can give many opportunities for ethical reflection.

NOTES

1. http://www.iris-sostenibilita.net/iris/docs/formazione/cfd2–2006/valutazione-finale-CFD-mag07.pdf
2. The ideas expressed by future teachers in relation to the question "what is an ecosystem, how do you define its boundaries?" are extremely varied and they offer numerous opportunities for discussion and further deepening of knowledge. Here are some examples:
   - *It is the totality of the animal and plant communities that occupy a particular area, e.g., the fluvial ecosystem (shoreline vegetation, macro-invertebrates, birds nesting along the shores . . .). The boundaries are given by the particular physical characteristics of that environment and that makes it different from other environments (i.e., lake ecosystems or the sea ecosystems etc.).*
   - *It is the web of relationships between the abiotic environment and the life forms living on it, organized according to more trophic levels. The boundaries can change depending on the phenomena that is being examined, i.e., a bush can be considered an ecosystem, but the planet Earth can also be considered an ecosystem.*
   - *It is the whole of the biotic and abiotic factors in a particular territory. An ecosystem has boundaries that depend on how much it has evolved to support the survival of each living being.*
3. Perhaps a more revealing approach is to compare the overall land claims between largely vegetarian and highly carnivorous societies. An overwhelmingly vegetarian diet produced by modern high-intensity cropping needs no more than 800 m^2 of arable land per capita. A fairly balanced Chinese diet of the late 1990s, containing less than 20 kg of meat, was produced from an average of 1100 m^2 per capita; the typical Western diet now claims up to 4000 m^2 per capita (Smil 2000). China's move to a higher meat diet impacts water security (Liu et al. 2008).
4. The energy problem cannot be dealt with separately from the problem of water: in the United States, for each kilowatt hour of supplied electricity, 8 L of water are consumed.
5. The truth resides in the very fact of the multiple viewpoints (Volk 1998).

BIBLIOGRAPHY

Aikenhead, G.S. 2006. *Science education for everyday life.* New York: Teachers College Press.

Aikenhead, G.S., and A.G. Ryan. 1992. Student's preconceptions about the epistemology of science. *Science Education* 76 (6): 559–80.

Arcà, M. 1993. *La cultura scientifica a scuola.* Milano: Franco Angeli.

————. 1995. La biologia come approccio alla complessità. In *Il senso di fare scienza*, ed. F. Alfieri, M. Arcà, and P. Guidoni, 467–99. Torino, Italy: IRRSAE Piemonte and Bollati Boringhieri.

Arcà, M., and P. Guidoni. 1987. *Guardare per sistemi, guardare per variabili.* Torino, Italy: Emme Edizioni.

Bagliani, M., G. Bravo, and S. Dalmazzone. 2008. A consumption-based approach to environmental Kuznets curves using the ecological footprint indicator. *Ecological Economics* 65: 650–61.

Ball, P. 2005. The earth moves most for humans. *Nature Digest* 2: 11–12.

Barbiero, G., E. Camino, and L. Colucci-Gray. 2006. Interplay Between Commons Boundaries and Webs in Natural Science. IASCP Europe Regional Meeting, Building the European Commons: From Open Fields to Open Source, Brescia, Italy, March 23–25.

Barker, J.A., G. Monger, I. Stevens, and T.J. King. 1986. *Biology Study Guide II—Part four: Ecology and evolution.* Revised Nuffield Advanced Science, Longman: Harlow.

Bateson, G. 1973. *Steps to an ecology of mind.* Winnipeg, CA: Paladin Books.

Bravo, G. 2004. Gaia, our new common. Some preliminary questions on earth system science and common-pool resources theory in the study of global human/environment relationships. Available online at http://dlc.dlib.indiana.edu/archive/00001347/00/Bravo_Gaia_040426_Paper120.pdf

Camino, E., G. Barbiero, and A. Benessia. 2007. Abitanti globalizzati e abitanti localizzati di un pianeta messo in crisi dagli umani. Cornice teorica e piste di ricerca didattica. *Azione Nonviolenta* 46 (8–9): 14–23.

Camino, E., G. Barbiero, A. Perazzone, and L. Colucci-Gray. 2005. Linking research and education to promote an integrated approach to sustainability. In *Handbook of sustainability research*, ed. W. L. Filho, 535–61. Frankfurt: Peter Lang Europaischer Verlag Wissenschaften.

Camino, E., A. Perazzone, F. Bertolino, and C. Vellano. 2002. A comparative analysis of various teaching approaches and different learning situations concerning the core concept of "ecosystem" in the Natural Sciences education. *Proceedings of the 2nd International Conference on Science Education*, Nicosia, November 11–13.

Capra, F. 2002. *The hidden connections: Integrating the biological, cognitive, and social dimensions of life into a science of sustainability.* New York: Doubleday.

————. 1997. *The web of life. A new synthesis of mind and matter.* London: Flamingo.

Cerutti, A. 2007. La prospettiva evoluzionistica nella formazione scientifica. La proposta della laurea magistrale EDeN (Evoluzione e Diversità dei sistemi Naturali). PhD diss., Univ. degli Studi di Torino.

Chambers, R. 1997. *Whose reality counts? Putting the first last.* Warwickshire, UK: Intermediate Technology Publications.

Chin, C., and L.G. Chia. 2005. Problem-based learning: Using ill-structured problems in biology project work. *Science Education* 90(1): 44–67.

Cini, M. 1990. *Trentatre variazioni su un tema.* Milano: Editori Riuniti.

————. 1994. *Un paradiso perduto.* Milano: Feltrinelli.

Cladinin, D.J., ed. 2006. *Handbook of narrative inquire.* London: Sage.

Colucci Gray, L., E. Camino, G. Barbiero, and D. Gray. 2006. From scientific literacy to sustainability literacy: An ecological framework for education. *Science Education* 90(2): 227–52.

Corn, M.L. 1993. Ecosystems, biomes, and watersheds: Definitions and use specialist in natural resources policy environment and natural resources policy division, July 14, 1993. Available online at www.cnie.org/nle/crsreports/biodiversity/biodv-6.cfm

Crick, F. 1958. Central dogma of molecular biology. *Nature* 227: 61–3.

Cutler, J., ed. 2004. *Encyclopedia of energy*. Oxford: Elsevier Science.

Desautels, J., and M. Larochelle. 1998. The epistemology of students: The "thingified" nature of scientific knowledge. In *International handbook of science education*, ed. B. J. Fraser and K. J. Tobin, 115–26. London: Kluwer Academic Publishers.

Dodman, M., E. Camino, and G. Barbiero. 2008. Language and science: products and processes of signification in the educational dialogue. *Journal of Science Communication* 7(3): A01. Also available online at http://jcom.sissa.it/archive/07/03/Jcom0703%282008%29A01/

Doménech, J., D. Gil-Pérez, A. Gras-Martí, J. Guisasola, J. Martínez-Torregrosa, J. Salinas, R. Trumper, P. Valdés. and A. Vilches. 2007. Teaching of energy issues: A debate proposal for a global reorientation. *Science & Education* 16(1): 43–64.

Eldredge, N., and S.J. Gould. 1972. Punctuated equilibria: An alternative to phyletic gradualism. In *Models in paleobiology*, ed. Schopf and T.J.M. Freeman, 82–115. San Francisco: Cooper & Co.

Fox Keller, E. 2000. *The century of the gene*. Cambridge, MA: Harvard Univ. Press.

Funtowicz, S., and J. Ravetz. 1999. Post-normal science: An insight now maturing. *Futures* 31(7): 641–6.

Funtowicz, S.O. 2001. Post-normal science. Science and governance under conditions of complexity. *Politeia* 62: 77–85.

Gallopin, G. 2004. Sustainable development: Epistemological challenges to science and technology. In *Sustainable development: Epistemological challenges to science and technology*, Santiago de Chile.

Gallopin, G., and H. Vessuri. 2006. Science for sustainable development. Articulating knowledges. In *Interfaces between science and society*, ed. A. G. Pereira, S. Guedes Vaz, and S. Tognetti. Sheffield, UK: Greenleaf Publishing.

Gallopin, G.C., S. Funtowicz, M. O'Connor, and J. Ravetz. 2001. Science for the 21st century. From social contract to the scientific core. *International Social Science Journal* 53 (168): 219–31.

Galtung, J. 1996. *Peace by peaceful means*. London: Sage Publications Ltd.

Gayford, C. 2004. A model for planning and evaluation of aspects of education for sustainability for students training to teach science in primary schools. *Environmental Education Research* 10 (2): 255–71.

Groode, T. 2006. Review of corn based ethanol energy use and greenhouse gas emissions. Working Paper 07–1, LFEE. Available online at http://web.mit.edu/newsoffice/2007/ethanol.html

Haberli, R., W. Grossenbaker Mansuy, and J. Thomson Klein. 2001. *Transdisciplinarity: Joint problem solving among science, technology and society*. Basel, Switzerland: Birkhauser.

Harding, S. 2006. *Animate Earth*. Devon, UK: Green Books.

Kincheloe, J., and C. Berry. 2004. *Rigour and complexity in educational research. Conceptualising the bricolage*. Buckingham: Open Univ. Press.

Jones, A. 2001. *Eating oil. Food supply in a changing climate*. London: Sustain—Elm Farm Research Centre.

Lakoff G., and M. Johnson. 1999. *Philosophy in the flesh*. New York: Basic Books.

Larochelle, M., and J. Désautels. 1991. "Of course, it's just obvious": Adolescents'ideas of scientific knowledge. *International Journal of Science Education* 13: 373–89.

Lederman, N.G. 1992. Students' and teachers' conceptions of nature of science: A review of the research. *Journal of Research in Science Teaching* 29: 331–59.

Linn, M.C., H.S. Lee, R. Tinker, F. Husic, and J.L. Chiu. 2006. Teaching and assessing knowledge integration in science. *Science* 313: 1049–50.

Liu, J., H. Yang, and H.H.G. Savenjie. 2008. China's move to higher meat diet hits water security. *Nature* 454: 397.

Longino, H.E. 2002. *The fate of knowledge.* Princeton, NJ: Princeton Univ. Press.

Manghi, S. 2004. *La Conoscenza Ecologica.* Milano: Raffaello Cortina Editore.

Marchetta, D. 2008. Formazione dei formatori alla sostenibilità. PhD diss., Univ. degli Studi di Torino.

Maturana, H., and F. Varela. 1998. *The tree of knowledge.* London: Shambala Publications.

Mayr, E. 1982. *The growth of biological thought: Diversity, evolution and inheritance.* Cambridge, MA: Harvard Univ. Press.

Mazzarello, P. 1999. *The hidden structure: A scientific biography of Camillo Golgi.* Oxford: Oxford Univ. Press.

McNiff, J. 2002. *Action research for professional development. Concise advice for action researchers.* Third edition available online at www.jeanmcniff.com/books/booklet1.html

Nabhan G.P., and S. Trimble. 1994. *The geography of childhood: Why children need wild places.* Boston: Beacon Press.

Novak J.D., and D.B. Gowin. 1984. *Learning how to learn.* Cambridge: Cambridge Univ. Press.

Odum, H.T. 1998. eMergy Evaluation. In *Energy flows in ecology and economy,* International Workshop on Advances in Energy Studies. Porto Venere, Italy.

Odum, E. P. 1997. *Ecology—A bridge between science and society.* London: Sinauer Associates.

Orr, D. 1992. *Ecological literacy.* Albany: SUNY Press.

Patfoort, P. 2006. *Difendersi senza aggredire. La potenza della nonviolenza.* Torino, Italy: Edizioni Gruppo Abele.

Patzek, T.W., and D. Pimentel. 2005. Comparison of incoming solar energy in the tropics with oil solar cells and several biofuel crops. Thermodynamics of energy production from biomass.*Critical Reviews in Plant Sciences.*

Post, D.M., M.W. Doyle, J.L. Sabo, and J.C. Finlay. 2007. The problem of boundaries in defining ecosystems: A potential landmine for uniting geomorphology and ecology. *Geomorphology* 89: 111–26.

Ravetz, J. 2005. *The no-nonsense guide to science.* London: New Internationalist & Verso.

Ross, K. 2000a. Matter and life—The cycling of materials. In *Science knowledge and the environment,* ed. M. Littledyke, K. Ross, and L. Lakin, 59–77. London: David Fulton Publishers.

———. 2000b. Energy and fuels. In *Science knowledge and the environment,* ed. M. Littledyke, K. Ross, and L. Lakin, 78–94. London: David Fulton Publishers.

Sachs, W. 1999. *Planet dialectics.* London: ZED Books.

Sarewitz, D. 2004. How science makes environmental controversies worse. *Environmental Science and Policy* 7: 385–403.

Smil, V. 2000. Energy in the twentieth century: Resources, conversions, costs, uses, and consequences. *Annual Review Energy Environment* 25: 21–51.

———. 2003. *Energy at the crossroad. Global perspectives and uncertainties.* London: MIT Press.

———. 2008. *Energy in nature and society. General energetics of complex systems.* Cambridge, MA: MIT Press.

Sobel, D. 1993. *Children's special places: Exploring the role of forts, dens and bushhouses in middle childhood.* Tucson, AZ: Zephyr Press.

Sterling, S. 2001. *Sustainable education, re-visioning learning and change.* Devon, UK: Green Books.

———. 2002. A baker's dozen—Towards changing our "loaf." *The Trumpeter* 18 (1), available online at http://trumpeter.athabascau.ca/index.php/trumpet/article/view/121/130

Thomashow, M. 1996. *Ecological identity*. Cambridge, MA: MIT Press.

Thomson, J. 1994. *Natural childhood*. New York: Simon & Schuster.

Tukker, A. 2008. Sustainability: A multi-interpretable notion. In *System innovation for sustainability 1. Perspectives on radical changes to sustainable consumption and production*, ed. A. Tukker et al. Sheffield, UK: Greenleaf Publishing.

Varela, F. J., E. Thompson, and E. Rosch. 1991. *The embodied mind: Cognitive science and human experience*. Cambridge, MA: MIT Press.

Virchow, R.C. 1858. *Die Cellularpathologie in ihrer Begründung auf physiologische und pathologische Gewebelehre*. Berlin: Hirschwald.

Volk T. 1998. *Gaia's body. Toward a physiology of Earth*. New York: Copernicus.

Wackernagel, M., and W. Rees. 1996. *Our ecological footprint: Reducing human impact on the Earth*. Gabriola Island: New Society Publishers.

Wittgenstein, L. 1969. *On certainty*. New York: Harper & Row.

7 Educating the Educators

Primary Teacher Education

Marta Angelotti, Anna Perazzone, Marco Davide Tonon and Fabrizio Bertolino with a contribution from Giuseppe Barbiero

EDITORS' INTRODUCTORY NOTES

Continuing on from Chapter 6, which was concerned with secondary teachers and specialists, this chapter looks at the experiences of teacher education at the primary level.

The authors start with acknowledging the impact that increasingly urbanized environments are having on the lives of children. Life is dominated by computer culture and virtual worlds, and play and movement tends to be rigidly structured. Experiences in nature are becoming less frequent for children, and so are their possibilities for autonomous movement (on foot, by bicycle), for choosing one's own friends and for freely organized play.

At the same time, few courses in environmental education appear in university curricula, and those that do are often considered as add-on courses, often of a normative and prescriptive character.

If a more ecological approach in education is needed, then teachers at the primary level have, perhaps more than at any other level, the possibility of developing a more holistic approach to learning in which science education can be an important tool in reconnecting people with nature. This raises important questions with regard to teachers' knowledge and their formative experiences. Drawing on the methodological foundations described in Chapter 6, this chapter describes activities in which the life and earth sciences can provide the context for raising the perception of oneself in relation to a deep interdependence with nature. Direct experiences in nature and a reflexive approach develop awareness of natural systems' capacity for self-regulation and of the multiple impacts of human actions. A reflective dialog is therefore initiated with beginning teachers on the potential for learning that is embedded in outdoor activities: these can become the nucleus for the organization of sustainable curriculum and pedagogy in primary education toward a scenario in which children can become effective constructors of sustainable worlds. A final section at the end of the chapter explores this point further by presenting the impact of activities of silence and meditation on children's biophilia.

INTRODUCTION

Loss of Ecological Identity and the New Needs of Children

What kind of environments and what experiences are essential for a complete and harmonious development of a child? How can we meet children's fundamental needs and help them to become part of a society and a fully formed person? What stimuli should a child receive in order to develop a harmonious and responsible relationship with the environment and grasp the essence of sustainable living?

Let's turn the question on its head and imagine what we would do if our purpose was that of preventing the child from developing, disconnecting him or her from reality, unable to act and completely deprived of those abilities for adaptation, expression, exploration and communication. Competences that are all potentially embedded in all human beings but that can only be expressed and become visible by means of an adequate interaction with the environment (Moore 1986).

First, we would have to put a limitation on the child's ability to move, go out, meet with peers, establish relationships and, further, deprive him or her of sound, light, objects and visual displays. In contrast, we could prevent the child from keeping still, by bombarding him or her with stimuli, keeping them in the constant presence of others and in continuous interaction, or more, we could make his or her environment uncertain, incomprehensible and inappropriately scaled, where each action is inappropriate and produces unpredictable responses (Lorenzo 2000).

It is then that we would find ourselves confronted with a "disconnected" child; one without ideas, apathetic or hyperactive, centered on him or herself and incapable of establishing a relationship with anybody other than themselves.

It is obvious that none of us would consciously desire to construct such contexts or even attempt to live in them. However, it is equally obvious that many of these situations, in more moderate forms, characterize the everyday life experiences of people, particularly those of the "small inhabitants of modern and wealthy cities of the West" (Tonucci 2002). It is indeed in the urban context that one of the many forms of structural violence of contemporary societies toward children becomes visible (Galtung 1969; Forni 2002). Children who are forced to live perpetually in an artificial environment where the complete scheduling of time and spaces generates stereotypical relationships; in which creativity, adventure and free play are sacrificed in the name of safety. The so-called bubble-wrap generation (Cadzow 2004; Malone 2007) is limited in the number of opportunities for autonomous movement and for socialization with other members of the community (Volpi 2004), and is also increasingly characterized by its life online. It communicates with others by means of chat-lines and blogs; it buys, sells and exchanges things on eBay; plays with the Playstation; meets with others on

MySpace; creates and passes time in Second Life. The native of the digital era (Prensky 2006) is in daily interaction with the virtual world: television, computers and videogames are a large part of the child's world. They provide symbolic and indirect experiences that are often so emotionally involving, pleasurable and interesting that they compete and displace the desire to go and immerse oneself in the "true reality" and meet "real objects."

In the era of media communication, thinking is reduced to what can be named. Direct experience is undervalued because the limitations of information delivery by the exclusive means of words, images or other forms of communication are not clear and apparent. As was mentioned in Chapter 4, verbal language has a tendency to permanently "fix" perceptions of reality in a reductive way. We become used to the process of "recognition" of objects, places, situations and problems (which are reduced to models) and to take on pre-established and stereotyped patterns of behavior. By such means we become part of a real process of "robotization" of the mind, as opposed to educating the child in the direct and "adventurous" discovery of the world (Ferrarotti 2005).

If, on the one hand, educational institutions are focused on allowing the child to be open toward the future, to be able to live in full the languages and experiences of current times, on the other hand, there is a pedagogical question of assessing the impacts of an increasingly rare (or sometimes completely missing) contact with reality, and particularly with the reality of natural environments (Louv 2005). Evidence of a lack of development of sensorial capacities (with consequent impoverishment of the human experience), difficulties of movement, obesity, increasing allergies and hyperactivity appear to be some of the signs of a disharmonic development that the child displays when deprived of experiences with nature (see the contribution by Giuseppe Barbiero, on p. 181, for more detail on this aspect). In fact, contact with nature can help the child to build his or her own ecological identity (Thomashow 1996). This comes with an increasing awareness of being part of an Earth system on which we are totally dependent for our survival, but that can also become a context for action to counterbalance the negative effects of living in the social environments of modern Western society.

In this regard, education, and particularly primary and nursery education—which are the contexts of our work—can play an important role. Teachers do not need new recipes or new programs of environmental education. What is needed is a renovated awareness of the new needs of children. It is then that direct experience will be prioritized and valued (also experience in natural contexts) and the knowledge to be promoted will be of a holistic and rounded kind, useful to generate critical inquiry and to "connect" people to reality.

The Context of Primary Teacher Education in Italy

Ten years ago, a formal training and qualification system for teachers was introduced at the higher education level in the Italian education system, consisting of either of a four-year undergraduate course for primary teachers, or a two-year postgraduate course for secondary teachers, very similar to that found in many other countries in Europe.

During these past ten years, we have reflected very deeply on the education of primary teachers, and particularly the role of science in preparing them to deal with issues of sustainability. One of the fundamental principles of primary-teacher education in Italy is to "promote students' participation, with the aim of achieving better consistency between the teaching methodologies and the formative objectives"(Ministerial Decree[1] 1998, article 2). It is a transformation of praxis that requires a reflection on the aims of education and contemporary needs. As was outlined in Chapters 5 and 6, systemic thinking, autonomy and critical thinking, sense of responsibility, solidarity and ecological identity are fundamental educational dimensions in a sustainable society. Even more so at the primary level, such dimensions should be pursued with the awareness of responding to the needs of a childhood that has profoundly changed in recent times.

One of the benefits of primary education is that, at least at the institutional level, it is not fragmented into disciplines. It is, therefore, an ideal place for developing new ways of educating and looking at reality, with approaches that cut across different realms of knowledge. In this respect, environmental education has often been seen as an appropriate means through which primary education addresses the needs of children.

A survey conducted in 2004/2005 looked at the range of courses called *Environmental Education* within the university curricula (Bertolinoet al. 2007a; Bertolino, Perazzone, Bertone 2007). Out of 277 degree courses, there were a total of 42 environmental education modules. Of these forty-two, eighteen were aimed at students of primary education. Thus, approximately half of the teaching in this area occurred at the primary level. On the one hand, this can be taken as a positive indication of the importance given to this topic, at least at the level of primary education; on the other hand, it calls for a reflection on the compartmentalization of environmental themes that still occurs.

In reality, courses and laboratories[2] in environmental education are the result of the efforts of only a few lecturers who, with some difficulty, have created opportunities for introducing the themes of sustainability. Although such topics have now become officially recognized by the academic world, their full incorporation and recognition is still a struggle. For those who are not fully immersed in it, environmental education is still perceived as an additional area of teaching, often of a normative and prescriptive nature, and not as a transversal dimension of global education. Hence, we are still far from Sterling's concept of *sustainable education* (Sterling 2001); and a "change of educational culture which can both develop and embody the theory and practice of sustainability in a critically conscious way" (28).

The research mentioned above (Bertolino et al. 2007a) also showed that those Italian university lecturers involved in environmental education belong to a variety of disciplinary sectors ranging from the natural sciences (biology, geology, forestry) to pedagogy and humanisitc sciences (anthropology, geography), often influencing one another, crossing into new areas that do not traditionally belong to the field in which they had originally trained.

In our opinion, this is a very positive trend, as we believe that by promoting and facilitating interdisciplinary communication, the aims of education can be redefined to address the urgent requirements of contemporary society, and the consequent needs of the children. The revolutionary idea of *sustainable education* (Sterling 2001) can be taken forward within the current educational system only by means of transdisciplinary visions. These originate from a convergence of objectives and a concerted use of interpretive schemes that are suitable for overcoming the cognitive obstacles preventing sustainable living.

Primary-Teacher Education at the University of Turin

It is on this convergence of perspectives that the Turin group has been working, so far, in the realm of science education (Perazzone et al. 2002). The theme of sustainability offers the common background for courses and laboratories in the natural sciences within the degree course in primary education. We are responsible for the courses in life sciences education and Earth sciences education, which include a small component of co-teaching, and are part of the first-year courses in the four-year program. In the fourth year, there is an additional course in natural sciences education and a few laboratories, which constitute the applied side of the lecture-based courses. Unfortunately, as we will explain later, interactive teaching is severely limited by student numbers, which can reach 150 (out of 571 students enrolled in 2008) for each lecture in the first year and about 70 students for the fourth-year course. The laboratories can host about 100 students divided into small groups, although there is a surplus of demand that cannot be met. In fact, students can attend laboratories of their choice from a fairly large selection, but for each one of those, there is a prescribed number of students who can attend.

In the following sections, we will explain in more detail the specificities of the courses and laboratories in relation to both content and methodologies with some examples.

Integration of Knowledge and Development of Conceptual Tools

The objective and subjective aspects of our ways of understanding the world are closely interwoven. Hence, how can we move within the complexity that characterizes life and the environments in which it develops? In order to make things clear, science tries to distinguish facts and phenomena by making them stand out from a background [. . .]. But how do we distinguish the events from their background? (Alfieri et al. 1995)

The reality that we observe is of multiple forms, consisting of intricate webs of interdependencies and a continuous flow of events, phenomena and processes that we can know only in part (Cini 1994). Each one of us believes that reality coincides with what we "naturally" perceive and we are not aware that we construct our reality from particular positions and points of view, both biological and cultural.

The choice between what is significant and what is negligible in the reality we observe does not only depend on the "facts" or what we are able to grasp through our senses (or the additional instruments we can use), but also on the cultural system and the observers' ways of thinking and their values. In this way, we can also say that the construction of scientific knowledge is both influenced by and has an influence on the cultural evolution of the ways of thinking about the world. We thus find ourselves in the condition of having to consider different levels of complexity and interaction: the complexity of the natural systems and our way of knowing and exploring a reality that is in continuous transformation (Maturana and Varela 1984).

Approaches to the Natural Sciences and their Integration

From this follows the need to put boundaries between events and background, and such boundaries can be located in:

- Space: when I choose to cast my attention on a detail—for example, when I take into consideration only the cell, an organism or a precise ecosystem.
- Time: because processes do not have a beginning and an end that are intrinsically defined, but they depend on the moment in which a time "zero" is being defined on the conventions, habits and the desire to simplify.

However, each space or time boundary corresponds to an abstraction that, on the one hand, allows us to divide reality (or at least one aspect) in sections and chunks that are more easily understandable; but on the other hand, it creates a separation from the context. Indeed, if the latter is not made explicit, it prevents us from having a holistic picture and making sense of the natural systems in which such portions or sections of reality are immersed.

One could say that, depending on the boundaries that are chosen, the construction of knowledge in the natural sciences is rooted in a series of different approaches; our reflection within the research group—as explained in Chapter 6—has led us to identify at least four approaches: structural, functional, historical and systemic.

As we have seen in Chapter 6, from a scientific point of view, there is equal recognition of the importance of such different approaches, which have undoubtedly contributed over the centuries to the construction of knowledge in the fields of biology and geology. However, from an educational point of view, it seems to us that the importance of the combined use of such approaches has not yet been fully understood. Such approaches are useful as tools for achieving an integrated view of reality and thus for making sense of notions and facts that are too often acquired as separate from an overarching framework (or context).

This consideration is even more significant from the students' point of view. Because they often miss the original context of reference and ignore the epistemological framework in which a particular form of knowledge developed (and this can be because it is implicit and/or because the holistic view is indeed missing), students are confronted with a fragmented scenario consisting of scattered information and disjointed ideas. These do not make for an articulated and holistic mental image of the object under study, with the result that too often scientific knowledge is perceived as something separate and different from one's own way of observing oneself and the surrounding world, and such knowledge is never integrated into everyday life. In this way, scientific knowledge and the idea of science that follows from this, tends to remain a "scholastic object," far removed from the learner and not an instrument for understanding what is around us.

Educational Aims: From the Contents of the Natural Sciences to the Idea of "Inclusion" and Sustainability

This discussion leads to the fundamental aim that shapes the planning of the educational activities introduced in the degree course for primary teachers: from an idea of knowledge as a "scholastic object" to an idea of knowledge that can include and orient us.

Seeing oneself as an isolated and "divided" organism within one's own body is an indication of a difficulty toward thinking in a systemic way. This difficulty can be seen in our students in relation to each of the hierarchical levels we refer to when we talk about life (from the cell to the ecosystem). One of the consequences is that of not perceiving oneself as included in the dynamics of the natural systems and therefore a belief of a strong man–nature divide. From this derives a worldview that is far removed from the idea of sustainability.

However, life and earth sciences can, indeed, provide the opportunity to perceive oneself in a relationship of close interdependency with nature. Consequently, this can contribute to building awareness of resource uses and the capacity of natural systems to self-regulate despite the excessive exploitation made by our species.

Starting from these premises, we have asked ourselves, as educators, a number of questions: How do we introduce disciplinary content and reflection about science to our students? How do we help them to build a holistic vision of the natural sciences through which they can develop an awareness of the fact that we are living on a finite and closely interconnected planet? How do we develop a sense of inclusion in the natural systems?

We need to think about *how* the knowledge of the natural sciences can become part of the knowledge of concrete experience: hence, which tools can be used to promote the integration of the different approaches and develop the perception (or sometimes the discovery!) of being part of the natural systems?

In the past few years, our group has identified and introduced the use of some *conceptual tools* (described in Chapter 6) as part of its educational practice. These are characterized by being adisciplinary (in other words, they do not belong to any particular discipline) and as such, they can be used as forms of interpretation of reality that are linked to generic ways of thinking. Conceptual tools can represent a kind of organizing thread, linking concepts that appear to be distant and unrelated but that can, in fact, be reorganized as part of the same reality, determined by the subjects and their various relationships.

In practice, we make use of a reflective teaching approach through which we can emphasize the structural function of the concepts in the life sciences and the earth sciences without encapsulating them within specific disciplinary sectors. Our intention is that of exploring such concepts by means of interpretive tools that can act simultaneously as links between disciplines, integration between the "two faces" of the natural sciences (life sciences and earth sciences), openness to and connection with environmental issues. Hence, what we seek to promote is a systemic cognitive style that is aimed at drawing the web of relationships that keep each living unit together and "uncovering the interconnections" that make systems, indeed, interdependent" (Mortari 2001).

The Peculiarities of the Educational Context of the Degree Course in Primary Education

From a theoretical point of view, the model used for designing our courses is very similar to the model described in Chapter 6, although there are inevitable differences in its applications.

This is largely due to differences in the typology of the students and their numbers. Chapter 6 refers to a limited cohort of postgraduate students (never more than twenty-five to thirty students for each course) holding a first degree and generally interested in their own disciplinary field. In contrast, for the degree course in primary education, numbers are much higher (from a few dozen in 1998 to 571 in 2008); such students often have limited disciplinary knowledge and are less keen to get involved.

Another substantial difference is the audience that such teachers will need to address: primary teachers will be required to teach all disciplines, and this can be both an advantage and a disadvantage The disadvantage derives from the fact that within the degree course there is a very limited number of hours dedicated to science preparation, and this is at odds with the level of preparation of the students who need to build up their knowledge to a level that is appropriate for the job of a primary teacher.

The advantage, however, is that they find themselves in a position where they can more easily build an interdisciplinary approach and appreciate the importance of integrating different knowledges in learning.

The combination of all such factors poses two kinds of problems. The high numbers of students make it impossible to make exclusive use of interactive teaching because it is necessary to adopt a lecture format. In addition, gaps in students' disciplinary knowledge need to be addressed, and this inevitably requires striking a balance between putting emphasis on content and looking at the methodological aspects of teaching. This is not unique to Italy and is, we believe, a common problem encountered in primary teacher education programs throughout the world.

Thus, how do we bridge such constraints with the need to build meanings and knowledge through the involvement of thoughts, feelings and actions (Novak 1998)? How do we stimulate that process of profound modification on one's own identity that is at the basis of any meaningful learning? The task is all but simple, and in the following sections we will try to explain our strategies, both in terms of choice of disciplinary content and methodological approaches.

The Content of the Courses

When focusing on courses, a key aspect to consider is the choice of content. This needs to include unifying and structuring concepts, which can be used to interpret and connect objects, events and processes, from different time and space scales and different points of view (Perazzone 2004). In substance, it is a matter of covering the main ideas of biology and earth sciences, promoting a reflective attitude and organizing such ideas within a complex and flexible conceptual web, which can be useful in different situations.

In the realm of the life sciences, for example, the sequence of topics we deal with starts from the exploration of the living organism and from its "definition" as a complex and open *system,* characterized by structural elements and relationships that allow for its maintenance.

Talking about living *systems* allows us to introduce the notion of different organizational levels and the continuous exchanges that exist between them. Moving from one organizational level to the next reflects the "choice" of defining boundaries that are continuously widened or narrowed down depending on the focus of the study.

In this regard, within the theme of the living systems, cell and organism are the easiest concepts to grasp, as they are thought of as objects that have well-defined structural boundaries (apart from functional ones). Again, this is both an advantage and a disadvantage. We are able to see boundaries when we examine something that appears to be well defined. However, we tend to focus our attention on what is inside and ignore the environment of which it is a part. In truth, one of the most essential characteristics of living systems is that of being open systems that continuously exchange energy,

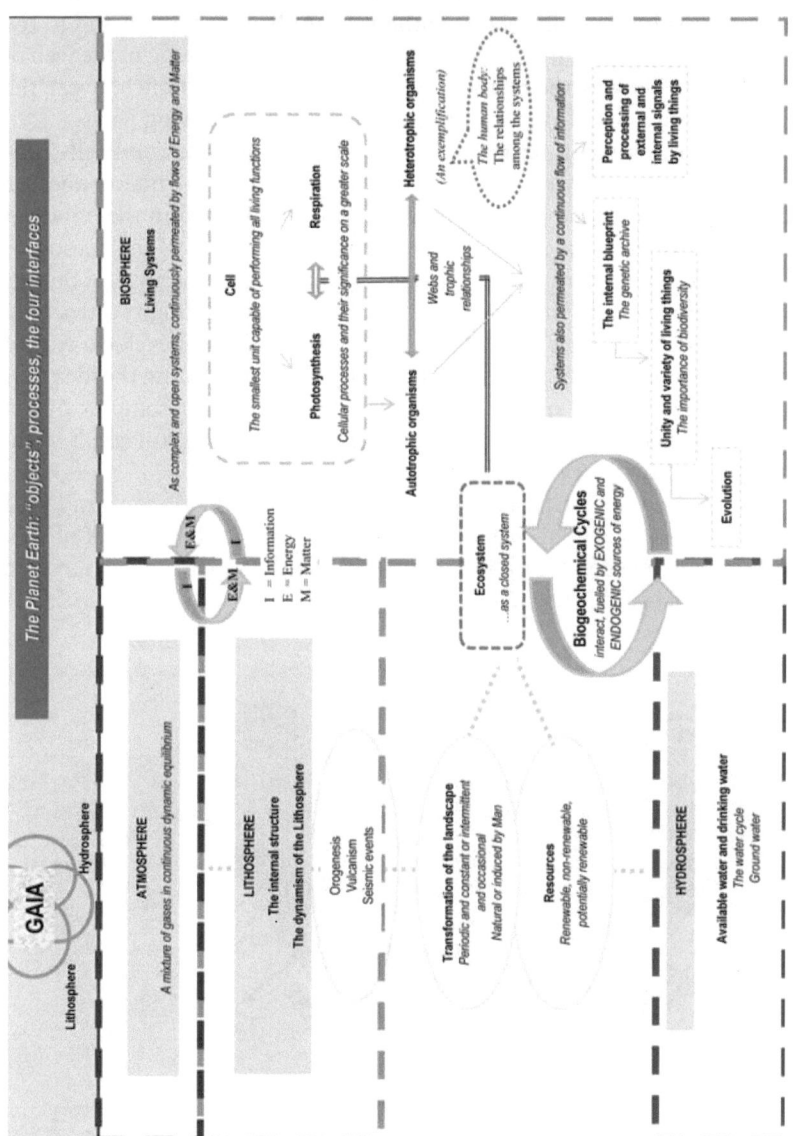

Figure 7.1 Integrated overview of the main topics covered in the life and earth sciences education courses (first year of the degree course in primary education).

matter and information with their environment. This awareness opens new avenues both toward the knowledge of oneself and the understanding of environmental problems.

The application of the conceptual tool *micro/macro* can help students grasp the importance of linking together and taking into consideration different dimensional scales. This can facilitate an understanding of processes that are often dealt with as part of confined orders of magnitude (molecular level, cellular level etc.). From this perspective, for example, we can compare and contrast cell respiration with photosynthesis, link them to the cell and to the concept of the ecosystem. In turn, this can promote the widening of the web of relationships, exchanges and flows of matter and energy, not only between the living components, but also by explicitly including the nonliving parts.

Indeed, by thinking in terms of bio–geo–chemical cycles and self-sufficiency of the ecosystems we can look at the planet Earth as an interconnected whole. The concept of the planet Earth is often well defined in the minds of the students because of its structural boundaries, but it is not understood or is little considered in terms of its exchanges of matter and energy. It is on this particular theme that we then focus our attention. Also, through co-teaching lessons, we endeavor to get students to reflect on the need to develop a way of looking at the world that goes beyond the disciplines, promoting the idea that the natural environment on which we all depend is such a complex object that no specialized disciplinary research can achieve any useful result without the help and support provided by other disciplines.

Indeed, leading the students to become aware that the planet is in fact a closed and self-sufficient system is the first step toward the concept of sustainability, which is dealt with more explicitly in the fourth-year course on "Natural Sciences Education."

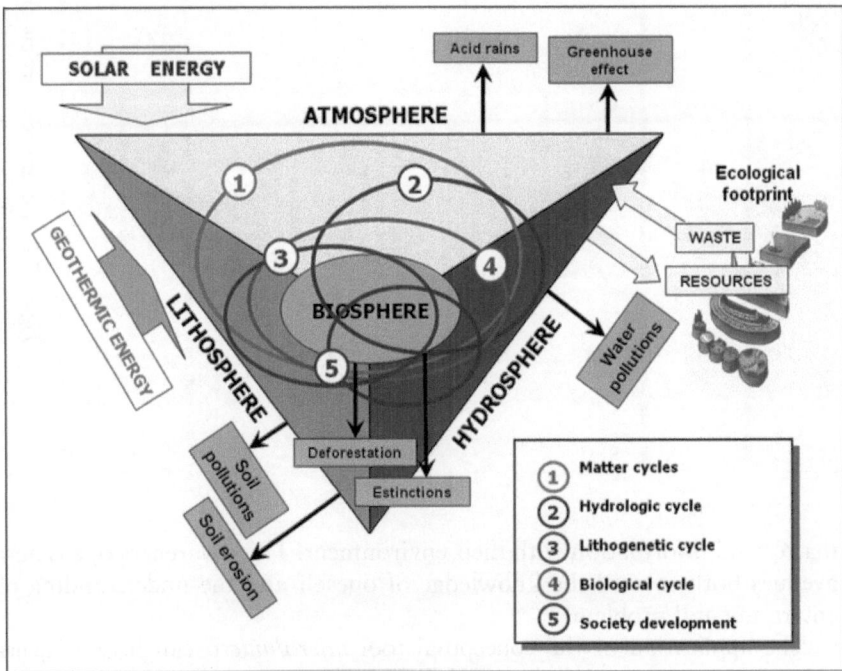

Figure 7.2   Overview of the fourth-year course.

This course represents a path of integrated learning about the planet. It focuses attention on the surfaces of contact (interfaces) through which all cycles of matter and energy flows occur, with particular attention given to the processes involving primary resources, such as air, water, rocks, soil and biodiversity. Then, the complex and systemic nature of interdependent relationships between living systems and physical context is recognized in the environments and situations of everyday life, at school and at home. Students can become aware of man-made modifications and system imbalances, which are at the source of great environmental problems (i.e., climate change, waste management and disposal, water and soil pollution). By means of examples of activities and educational reflections related to such themes, we demonstrate the possibility of becoming responsible actors who are conscious of material and energy resources in a scenario of sustainable development.

Methodological Aspects

We have tried very briefly to present part of the content that we present to our students; however, in the light of the problems indicated previously, what kind of methodologies do we put in place? Which tools can we use to stimulate reflection both at the level of the disciplines and at the level of teaching implementation?

While a transmissive/lecture style mode is privileged (for the reasons already expressed above), some individual and group activities are nevertheless introduced. These are aimed at making students' preconceptions emerge, the kind of approaches they use to deal with particular topics, the relationships between the different topics that are being presented and the possible conceptual obstacles. Students' products are then collected lesson by lesson, analyzed and re-presented in subsequent lectures, not for the purpose of assessment, but for triggering students' reflections on stereotypes and misconceptions that often prevent them from reaching a wider understanding. As we mentioned earlier, knowledge without reflection tends to remain a scholastic object rather than an instrument to include and orient the learner.

For example, by asking students to draw the Earth's internal structure we can illuminate the common misconceptions that make people think that the magma would come straight from the center of the Earth; that is, its nucleus. This misconception is generated by the classic conceptual obstacle linked to the invisibility and inaccessibility of geological structures (Ferrero and Gimigliano 2007).

More significantly is to ask the students to draw a volcano. In almost all cases, they draw a conic structure with a circular whole on the top. This partial and stereotypical view does not take into account that the majority of volcanoes on the Earth are actually underwater (hence invisible), and are constituted by wide, tabular structures with linear fissures; alternatively, if the volcanoes are emerged, their structures can have different shapes (shield volcanoes, calderas, domes, necks).

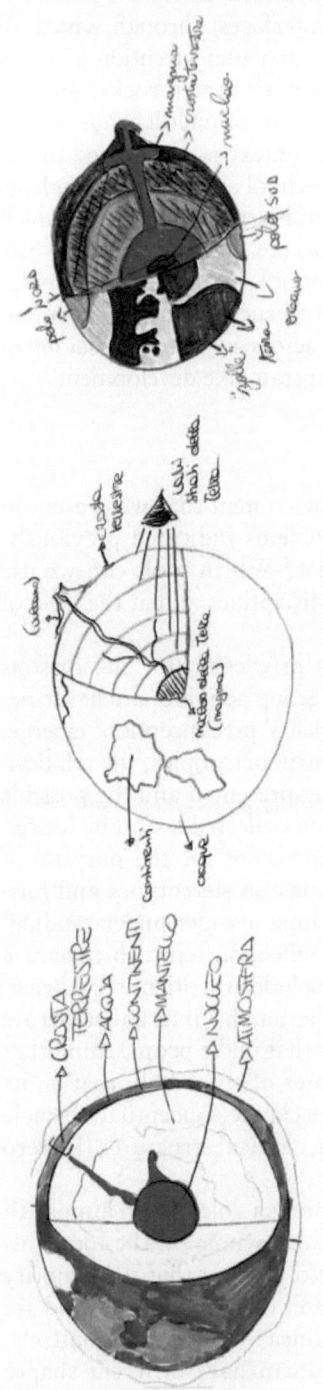

Figure 7.3 Example of drawings of the internal structure of the Earth presenting a magma-centric view (held by about 18% of the students).

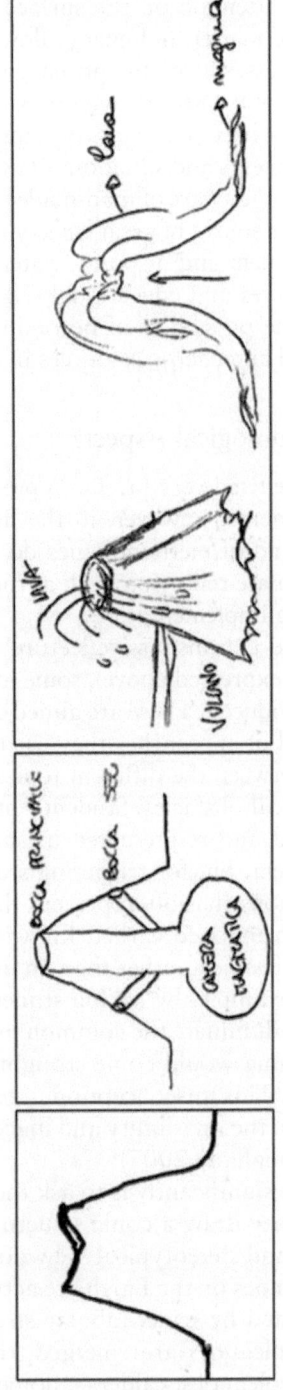

Figure 7.4 Example of drawings of the structure of a volcano, with the dominant conic type (held by about 95% of the students).

Another example derived from the life sciences can help us to clarify even better the need for weaving together the different approaches (described on page 159 in this chapter and on page 127 in Chapter 6). For the purpose of helping students handle the cognitive shift from the cellular level to the organism (see Figure 7.5.), we ask them to describe and draw the journey of a chunk of food inside a human figure. Both the many literature sources (Alfieri et al. 1995; Arcà 2005; Carvalho et al. 2004; Clément P. 2001; Giordan and De Vecchi 1990; Psarros and Stavridou 2001; Reiss et al. 2002) and the drawings we have collected over the years (Perazzone 2004) allow us to claim that such a journey is usually represented through a structural approach. Sometimes this is complemented in the description by a functional approach that only considers the digestive tract, however, hence losing sight of the process as a whole. In the majority of cases, such representations neither allow for imagining the dynamics of physiological aspects nor the importance of the interrelations between different organs. What never comes through, neither graphically nor verbally, is the need to connect what happens at the macroscopic level (food ingestion) with what happens at the cellular and molecular levels.

The anatomical and physiological study of the human body is all very well. However, if this is not integrated at any point with a systemic approach, it will be very difficult for students to understand that, ultimately, the digestive, circulatory and respiratory systems all have one common function: that of providing nourishment, in the form of nutrients, water and oxygen,

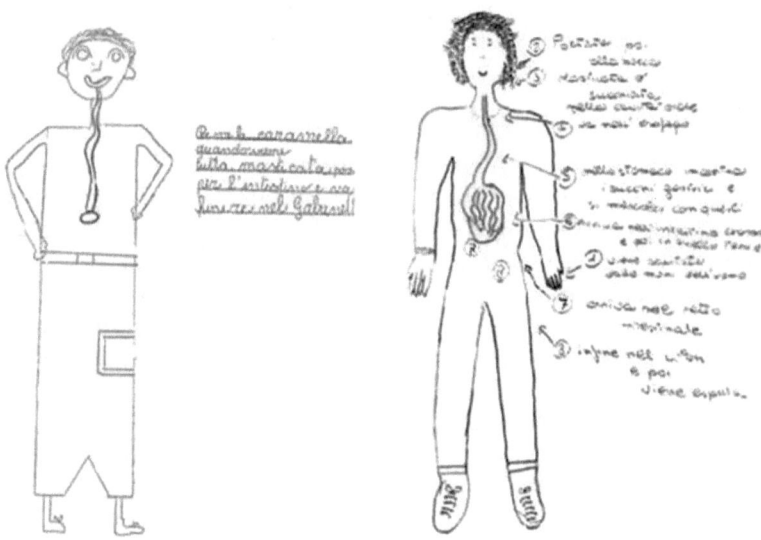

Figure 7.5 Examples of mental representations held by children and university students "Represent the journey of a sweetie inside your body."

to the parts of a body, which, just like the body of all living beings, needs energy and it is permeated by a flow of matter.

Presenting students with their own works often has a very strong impact on them, particularly if their drawings are juxtaposed with those of primary school children, in which it is possible to observe similar difficulties. Misconceptions and partial representations are sometimes more obvious in children's works, but this is only because they are expressed in a naive language that is less rich in scientific terms, behind which the adults often hide. It is through this dialog and reflection that students become aware of their own, and other peoples', conceptual obstacles and the educational importance of making such obstacles emerge for the purpose of overcoming them.

On the left is the drawing of a child in primary 4; on the right, the drawing of a first-year university student in primary education. In both cases, the sweet is not assimilated by the body.

About 30% of university students do not talk of substance absorption; 45% talk about absorption or assimilation and 20% link the digestive system with the circulatory systems. Less than 10% of the students connect the macroscopic level with the microscopic level by talking about cells (student number 82, 2003/2004 academic year).

To stimulate students to get involved and reflect on their process of knowledge acquisition has further educational value for us because it provides students with an educational model that they can reproduce in schools.

THE POTENTIALITIES OF A DIRECT RELATIONSHIP WITH THE NATURAL ENVIRONMENT

The Educational Excursion in Nature

An outdoor trip is always a fundamental formative event, for students as well as tutors. Direct experiences are, in fact, central to cognitive processes, especially if they constitute moments of problematization (Battistini et al. 1983; Jaén 2000), observation and analysis of real situations, as well as review of and reflection on the educational path made. It is only through direct field inquiry that students can visualize and appreciate the three-dimensional structure of natural objects, assess their proportions in space and see the result of the complex dynamics of natural processes. The natural sciences are, in fact, often characterized by conceptual obstacles that derive from the inferential nature of most reasoning and knowledge construction in this realm (Battistini et al. 1982). Such obstacles can interfere with the learning process, generating mental representations of reality that are dissimilar from scientific ones. Direct field work provides opportunities to raise awareness of such difficulties and the means to overcome them (Massa and Pedemonte 1998).

In addition, by promoting an empirical and holistic approach to complex systems, it is possible to facilitate the hard task of discerning the numerous variables that have interacted, or that are still shaping a particular event, observed or inferred from the collection of cues and evidence. No single sample collection (i.e., rocks, fossils, animals, plants etc.), model or textbook can ever represent equally well the wholeness and specificity of the object contextualized in its original environment or linked to the event that generated and transformed it. The direct experience of inquiry in the field promotes several research skills that cannot be reproduced by means of simulation in the laboratory, nor by means of a virtual exercise at the computer. The direct immediate experience in an environment provides a strong motivation for learning because it confronts the learner with concrete problems.

Apart from the cognitive facilitation, there are also the many opportunities created by the different educational environments. The natural environment, without the physical barriers that confine students in small spaces, promotes the development of a new relationship between students and tutors that is more focused on the communal sharing of the experience. Often, in fact, fieldwork is carried out in small groups, and this creates new opportunities for creating knowledge and socializing with peers and the teacher as well. This results in a deepening of interpersonal relationships in a context that is not as rigidly formalized as that of school (Garcia de la Torre 1994). The need to formulate hypotheses directly and communicate the results from the field inquiry requires an exchange of information and represents a primary factor in supporting socialization within and between the groups.

Each student who is working on a team with peers is encouraged to solve problems by means of active involvement. He or she can feel the need to activate his or her own manual abilities in order to get information and data and to use these to formulate and verify hypotheses, as well as choosing appropriate communication techniques for sharing his or her experiences with the other members of the group.

This kind of shared experience develops self-confidence and trust in the group and makes students more responsible and aware of their own learning process. Moreover, if this approach can be turned into a habit, it would reinforce the interest and desire for discovery of natural phenomena that can be observed in everyday life.

Essentially, an excursion is a good opportunity for applying what is termed a "differentiated pedagogy" (Colombies 1997). The different educational approaches required for carrying out the various practical activities in the environment can engage students at different levels, involving the cognitive as well as the emotional, relational and motor spheres (AA.VV 1998).

The *cognitive sphere* is engaged by a practical–observational approach, aimed at developing empirical knowledge, based on observation and direct manipulation. On the one hand, this leads the students to the practical discovery of the diverse natural components and the relationships between them. On the other hand, it leads them toward a reflection on their own relationship with the environment and the impacts of their actions. This reflection gives pupils a sense of civic responsibility, developing their ecological consciousness.

The *socioemotional sphere* refers to a sensorial and esthetical contact: the involvement of one's own senses and the release of emotional reactions. Quite often, coming in contact with the surrounding landscape stimulates curiosity and contributes to the construction of a relationship between the environment and one's own personality.

Finally, there is a *motor sphere,* which is often overlooked and traditionally separated from the other two. An excursion is a time for walking, often on difficult terrain. This is an opportunity for children to learn to make secure steps on the ground, using appropriate techniques and learning to optimize their physical energies or to adapt to the different climatic conditions. This contributes to the development of abilities for autonomous movement, assessment of the difficulties and the possible dangers.

Fieldwork activities allow for a perfect integration of the three spheres just described, making learning truly meaningful (Novak 1998). Confronting and overcoming small difficulties on terrain strengthens the child's self-confidence and it often triggers positive emotions and a desire for knowing and discovering new things. To reach the top of a hill or a mountain means to have achieved a lofty target, both materially and symbolically, a target that is also connected to the feeling of being an integral part of the surrounding nature.

An excursion, therefore, is a precious opportunity for establishing direct contact with the environment and experimenting with different modes of relating to it: from moving to observing, listening, smelling, taking pictures, drawing, gathering objects to be used in class and interpreting (Ferrero and Tonon 1992). This contact with the natural and human environments offers students a variety of stimuli, sensations and emotions that could not have been thought of a priori; at the same time, however, such activities allow them to simulate a process of research and experimentation, developing the ability to infer starting from observation and data gathering (Ferrero, Provera and Tonon 2004).

The Educational Context of the Laboratories

The experience of designing laboratories for trainee primary teachers at Turin University has shown that fieldwork is an excellent opportunity for illustrating the scientific methodologies of inquiry by means of guided, systematic and integrated observation, and for developing an understanding of, and reflection on, complex biological and geological phenomena (Tonon 2004). The concepts that are proposed in the traditional context of the lecture are revisited in the context of the laboratory, and this allows for an integration of the educational experience on at least three levels.

In the first instance, there is an opportunity to apply theoretical concepts to real contexts and objects, making what was said earlier about the conceptual obstacles experienced by both children and adults in the process of knowledge construction in the natural sciences real for the students.

Secondly, the context of the laboratory can expand opportunities for inter-disciplinary connections because the various activities are conducted in co-teaching (with a lecturer of life sciences and a lecturer of earth sciences) and also because it is possible to deal with specific themes, which were deliberately chosen for the purpose of showing the need for interdisciplinary integration.

Finally, it is possible to balance the relationship between lecturers and students. The laboratories are open to a maximum of thirty students who can choose a specific laboratory from a broader selection of courses. This means that students are usually more motivated and attend regularly. This is important because, for this part of the course, there is no assessment. Hence, positive relationships and motivation make for productive communication and for a more gratifying experience for all.

Going back to the relational aspects, another feature of the laboratory is the residential experience. Lecturers and students live together for a period of two to three days and this generates a significant change in group dynamics; the possibility of sharing communal times and spaces, beyond simple teaching encounters, allows for a more relaxed and less formal atmosphere. Hence, there is much more motivation and participation in the proposed activities. This compensates for, in many respects, the limited interaction that is possible between student and lecturer, and among students themselves, in a lecture-style course structure.

The relevance of what is being said here is testified by the voices of the students who, at the end of each laboratory, are asked to express their personal comments on the positive and negative aspects of their lived experience:

> . . . Ejected and thrown from Internet to the raw material. The pleasure of the discovery experienced in the real context. The relationship of "friendship" between teacher and student. The sharing of this experience with other participants in the laboratory. The beautiful approach with nature.
>
> . . . I really did not believe I could have so much fun and feel so good. This laboratory is the proof that it is possible to have a relationship with the lecturers and it is really beautiful indeed! Given the students' numbers in Primary Education, to be able to talk a little with the lecturers and have a dialog appears impossible, a utopia. But it is not. And this makes me happy . . .
>
> . . . The laboratory was very interesting, beyond my expectations. I really liked the atmosphere between us. The activities were engaging and it was good to be able to share our experiences, emotions . . .

In the following sections, we will briefly present three laboratories that include a large component of fieldwork. Two of these laboratories included a residential phase in an alpine refuge, whereas the third laboratory consisted of excursions of a day or half a day. Overall, the educational experience presented here covers the variety of approaches and aims that we are trying to introduce, both from the point of view of content and methodologies.

Laboratory "Discovering the Natural Environment through the Educational Excursion"

This laboratory offers the opportunity to have an experience of direct contact with the natural environment in order to explore its constitutive elements and grasp the complexity of the relationships within it. Of the three laboratories described here, this one is strongly focused on the cognitive sphere and is more directly linked to the topics of the first-year courses (life science education and earth science education).

The main objectives can be summarized as follows:

- To experience modes of interaction with the environment that can help students to recognize the processes for observing, investigating and interpreting reality (Mortari 2001);
- To develop some conceptual organizers of the life sciences and the earth sciences: starting from the complexity of the observed and inferred phenomena, the dimensional relationships of the objects of study and their dynamic evolution, linked to geological and biological times;
- To apply theoretical knowledge previously acquired in the context of everyday life, with the purpose of interpreting the complex reality of the regional territory in which we live;
- To reflect on the concept of environment as a complex and dynamic system in which "natural" components (flows of matter and energy) are closely interwoven and interacting with social and cultural components.

From a methodological point of view, a deliberate choice was made of privileging aspects of learning through perception, senses and play because of their particularly formative value for future primary teachers. Students can experience the activities themselves and easily reproduce them with their own pupils. In the course of the activities, they can overcome different kinds of difficulties, some that are more obvious and others that are unexpected. Unusual experiences, comparing personal reactions, assessing the potentialities of the chosen site, experiencing the difficulties of organizing and managing the activities, developing the strategies for improving the educational potential—all these allow future teachers to elaborate on their own outdoor experiences with primary pupils.

Some of the tasks chosen are inspired by classical environmental education activities (Borgarello and Trusel 1991; Borgarello et al. 1997; Cornell 1998). Simulation games and sensorial perception, followed by a metareflection on the experiences, are a particularly integral part of the activity. The game of the *Artificial Path* (Borgarello and Trusel 1991), described in Box 7.1., is an example of such an activity.

Box 7.1   The artificial path. (Borgarello and Trusel 1991)

A short path of about 100 m within a natural environment is identified. The game leader places some "artificial" objects along the path, which students try to locate and record, walking in single file, without talking. At the end of the walk, the various records are counted to indicate the number of the objects that have been identified. Invariably, the maximum number recorded is far from the actual number of objects that was placed—and not hidden—on the path. The students are taken along the path again and the objects are revealed. This is followed by a moment of collective reflection with the purpose of analyzing the possible reasons that could explain such a discrepancy in the individual results. Although it was apparently simple, the task was deliberately ambiguous, and this concealed a series of difficulties that inevitably affected the final results. For example:

- The difficulty of distinguishing what is "artificial" from what is considered "natural." The term *artificial* is often and most commonly associated with a man-made product. However, something that the observer considers to be likely and plausible to be found in that context might be considered "natural." For example, a form of waste in a wood can be natural, while a volcanic rock placed on a mudflat may not be. A snail shell pierced on a stick may be artificial but a pine cone stuck on a trunk by a woodpecker may not be. The observer's knowledge greatly influences the development of the game.
- The difficulty of perceiving objects because of their high mimetic potential. The mimetic characteristic of a product is not an absolute value, but it is determined by the context in which the object is located and also by the observer's perceptive abilities.
- The difficulty of individual expectations. The ambiguity of the task neither provides information about the type of objects nor about the modalities of their placement. This determines expectations that lead to research expectations that are often predictable by the game organizer but that are not always productive (ratio of expectations to results).
- The difficulties concerning the limitations of the observer. The motor activity and the acoustic, visual and postural information that is perceived inevitably channels the observer's attention in particular directions.

Such considerations lead to the understanding that each individual applies a particular scheme of perception and organization of information. This is used more or less consciously and leads to a selection of different elements. What is "evident" for some people may not be for other people. People's mental schemes originate from a combination of factors related to multiple components (emotional, cognitive, sensorial). The possible variation of environmental conditions

(continued)

Box 7.1 *(continued)*

(luminosity, temperature, climatic conditions . . .) and/or psychological conditions (anxiousness/serenity, calm/preoccupation, interest/boredom, happiness/unhappiness . . .) influences the perception of an environment, making it different not only for different individuals, but also for the same person in different situations. Hence, the perception of an object becomes a subjective, and not an objective, factor. This is linked to the interaction between the observer and the object or between ourselves, our knowledge processes and the environmental reality. An environment that is independent from our way of perceiving and knowing does not exist, and in this way, knowing the environment can lead to knowledge of ourselves.

Other activities are more specifically targeted on the disciplinary aspects of the natural sciences, although the game-based element is kept as part of the methodological asset.

Box 7.2 Chart of sounds and pictures.

On a plain field, students are divided into small groups. They have to identify an area of their own choice, measuring approximately twenty steps by twenty steps. They represent such an area on a piece of cardboard, including the cardinal points and the location of various environmental components (trees, bushes, herbs, rocks, animal tracks etc . . .). This phase of the activity is called *territorial mapping*. Subsequently, each group completes the chart with a corresponding sound map. In order to do this, each member locates oneself at each cardinal point, focusing attention and trying to identify the sounds coming from both the inside and outside of their square (see Figure 7.6).

This task required students to apply knowledge and abilities previously acquired during a cartographic laboratory. Potential difficulties concern the orientation of the map, the appropriate choice of scale, the modalities of representation and classification of the objects. The activity requires a first phase of data recording and collection of samples (trees, bushes, animals etc . . .), followed by a phase of discussion and comparisons of notes and samples with the information contained in the texts. This task allows students to learn about the parameters that are useful for identifying each object (What is the height of the trunk? The shape of the leaves? The appearance of the fruits? The color of the bark?) and to practice using the keys, looking for the interpretive keys, the explanatory power and good and bad aspects, particularly in relation to the users (specialists, adults, children).

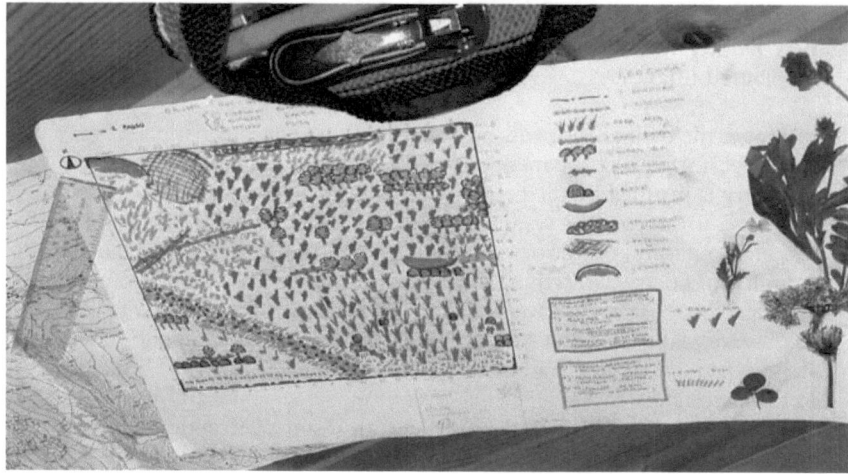

Figure 7.6 Example of a map of a territory of about 400 m².

Laboratory "Us and Nature"

The laboratory *Us and Nature* offers students the possibility of investigating their own relationship with nature in a personal fashion, with a view of achieving both personal and professional growth. By means of practical individual activities, discussions and group activities, students are introduced to the following themes:

- The psychology of landscape (Ammaniti and Stern 1991; Carbaugh 1996; Gallo Barbisio 2001, 2002) and particularly the role of the external environment in the construction of personal identity.
- The value of direct experience in constructing knowledge. The observation of the environment and the reflection on the time and space that characterize the experience.
- The man–nature relationship, lived and narrated in the first person through theatrical and artistic techniques, poetry, as tools for reflecting on one's own relationship, past and present, with nature.

With particular reference to the latter, we describe here an activity that leads to extremely gratifying products, which well exemplify a process of re-discovery of our ecological identity (Thomashow 1996).

Laboratory "Natural Resources"

This laboratory is designed for students of early years education, for whom there is often less choice of targeted laboratories. In fact, this laboratory emphasizes the use of manipulation practices and artistic expression for the creation of artifacts using natural primary materials previously collected

Box 7.3   The use of the haiku poetry: verbal pictures and the associated images.

In Japan, the haiku represents an important and characteristic part of the deepest nature of Japanese culture. The educational use of this type of poetry is justified by the belief that it is difficult, by means of common language, to capture the systemic organization of nature within a holistic vision. No natural object or event is too banal or small to be excluded from haiku poetry. Everything, in fact, can provide a series of stimuli that can feed the sense of discovery and re-discovery, provided one is able to observe reality through "fresh eyes," freeing oneself from one's own stereotyped convictions and one's preconceptions to touch the boundaries of the rational.

Each haiku is a little defined universe, an instant perception of time and space (Vasio 1999). Because each event is natural and dynamic— "alive"—the process of composing a haiku, even in its simplicity, allows such movement to be expressed. By means of word associations and the resulting metaphors, the poem illustrates the objective and emotional relationship in the human–nature interaction.

Although the haiku is a vehicle of this communion, the poem is never a simple, realistic description but also a verbal metaphor that always needs to be interpreted as part of a worldview that goes beyond conventional schemes. Matsuo Basho, one of the highest haiku poets of middle 600 said, after he read a composition by one of his students: "your weakness is that of wanting to impress. You devise splendid verses to describe faraway things; you should do this for those things which are near to you." In haiku poetry, the writer is required to express nature in its wholeness and complexity, but without indulging in its exotic and wild character: water, rocks, flowers, sun, clouds and stars, animals, plants, sea and wind and, along with all this, the observer's emotions. This includes the pain and the joy of mankind, the ecstasies and the torment.

The haiku composition technique is important to us because of the emotional component and the relevance of subjective experience, making this activity an important aspect of the exploration of the relationship that each one of us establishes with the natural environment. In our activities, students are asked to find a quiet spot in the woods and observe some natural elements or events that are happening around them. The task is that of capturing what is being observed through the writing of a haiku, which can encapsulate the students' ways of looking at nature (Bogina and Roberts 2005). Afterwards, each poem is transferred onto pieces of paper of equal size, kept anonymous and placed in a basket. In small groups, students equipped with cameras take out a few random poems from the basket and they spread themselves out on the field to capture by visual means what is verbally expressed in the haikus. For the purpose of sharing the interpretations, each haiku (the "verbal" pictures) are associated with their correspondent visual pictures: a collective reading

(continued)

Box 7.3 *(continued)*

of the works is then carried out by means of video projection. The final art product usually produces positive emotions, often very strong ones, which are shared with great satisfaction by the participants. The strange mirroring game that is created between the emotions that the writers have experienced while in nature and the interpretations of the subjective experiences described in the written words given by other people illustrates both the diversity of each person's way of being in the world and the communal perception of being part of one ecosystem. In the next section, we reported two examples of associations between haiku and pictures:

Flimsy drop
run over green trajectories
measuring time

A wooden web
of ancient roots reminds me
of the bonds with the earth

Plate 7.1   (a) and (b) Examples of haikus and associated pictures.

during outdoor fieldtrips. The main objective is that of uncovering the relationships between natural resources and objects of everyday use.

Classroom activities are alternated with thematic excursions. In this way, students can first engage in the collection and organization of the materials and subsequently manipulate such materials, breaking and recombining them, to create products expressing the close connection between humans and nature.

Box 7.4   Food resources.

Working on primary resources means reflecting on the fundamental needs for human survival, and such survival depends on nature. The laboratory is centered on the natural and ancestral need of gathering from the wild in order to eat, dress, build shelters and hence to sustain one's psycho-physical existence.

The primary constituting aim of the activity was to enter into contact with the essential spirit of self-sustenance and pure physical survival.

Direct gathering of materials was a central part: it required total physical involvement and stimulation of all senses for those who took part in the activity. To walk along and through different territories and environments (meadows, fields and woodlands) was directed by a precise aim. In this case, exploring and perceiving were not vague actions, but they were guided by the will of finding any kind of natural and edible products. Searching in the soil and among the vegetation to recognize the species or materials that were found; removing, uprooting or cutting the material; smelling it and sometimes having a taste of it were the series of actions that preceded the gathering of the materials for oneself.

All the natural materials that were gathered and identified as being edible underwent some kind of processing. The different species and the edible parts were selected out, cleaned and cut in order to be cooked according to their properties. All this went into the lunch that was provided for all participants.

Hence, by means of the involvement of the entire sensory experience, students managed to rediscover a direct relationship with nature, rebuilding that ancestral link that has always been fundamental to the survival of mankind.

Here is what one of the students said:

> Taking part in this laboratory was a bit like coming back to the times where each simple little thing could generate great enthusiasm. More specifically I would like to refer to the first excursion [. . .], because I had never had an experience of gathering food in the wild. This excursion was useful to me because I have seen and participated in an activity which can be repeated with the children, but also at a personal level, because after the collection and preparation of the various types of vegetables, I have learned to appreciate more the various plants and vegetables that there are. Now, in addition to eating a wider variety of vegetables than I used to do before, every time I find myself in a field or meadow or when I am in the car waiting at the red light in a tree-lined avenue, I spontaneously turn my head to look for species I can recognize.

Plate 7.2 Collective action of identification and collection of wild plant species.

Students' reflections start from the realization that the human species has always needed primary resources for its survival: starting from the need for nutrition, finding shelter and covering the body. Hence, the main three themes that are dealt with are food resources, the natural fibers—which are used for manufacturing textiles—and building materials. All of this is rediscovered and appreciated in the light of their use for satisfying primary needs.

For example, the path of production from an item of clothing to the original natural fiber was experimentally reproduced, and vice versa, a woven fiber was obtained from a piece of raw material (nettle, linen, canopy). In other excursions we have gathered and selected stone and wood materials for the construction of miniature models of houses and other artistic products that represented that particular territory. In relation to food resources, we selected, gathered and prepared some wild herbs until we had a complete meal, which was then consumed by the participants in proximity of the site of collection.

CLOSING REMARKS

As mentioned at the beginning of this chapter, in order to respond to the needs of new generations and to begin to create sustainable societies, there is a need for building knowledge starting from direct experience. It is indeed in the relationship between physical and psychological reality that the complexity we have to learn to explore resides. The desire to explore, the projection toward continuous forms of research, can develop

Plate 7.3   Selection and cleaning of collected species, prior to cooking and consumption.

and become autonomous only if the relationship with reality takes place through interest, curiosity and positive emotions. In all this, the natural environment plays a crucial role, as does scientific knowledge: indeed, by virtue of being directly applied to the real world, such knowledge becomes readily interwoven with all other fields of knowledge. Thus, at this point, the importance of preserving the widest possible interdisciplinarity becomes clear. At the school level, in fact, it is necessary to offer a broad and interconnected view that is not too dense and formalized; a dynamic picture that can allow for the formulation of new questions and inquiry, and through which the web of knowledge can be enriched with new links and interconnections. "We open a door and we step inside," as remarked by Longo (1998). In contrast, other school levels are directed toward the deepening of disciplinary knowledge, which, without an overarching frame, can become meaningless. It is for this reason that we consider interdisciplinarity and systemic approaches fundamental elements of primary education: if children learn to link together and connect, if they learn to build webs, albeit imperfect ones, they will acquire competences that will be difficult to develop at the school levels that follow. So, when we talk about social and environmental sustainability, we believe that the relevance and significance of such competences does not need to be argued any further!

REVEALING CHILDREN'S BIOPHILIA

Giuseppe Barbiero

While the relationship with nature is recognized as one of the most important components of human beings' physical, psychological and moral development (Kellert 1997; Camino 2005), our scientific knowledge about nature's impact on the different phases of development of the child, adolescent and adult is rather limited.

There is a great deal of anecdotal knowledge confirming the importance of this relationship; however, we neither have a systematic analysis nor a coherent theoretical framework for understanding the role of nature in child's development (Kahn and Kellert 2002). At the end of the last century, Edward O. Wilson put forward the hypothesis of the existence of a human instinct that he called biophilia (Wilson 1984). The biophilia hypothesis offers an evolutionary explanation for the close man–nature relationship (Kellert and Wilson 1993), and over time, a number of empirical proofs have been gathered to support it, so that currently, the biophilia hypothesis "can provide a unifying framework across numerous disciplines to investigate the human relationship with nature" (Kahn 1999), and it can be reasonably considered a plausible evolutionary explanation for a series of innate human behaviors of relationship with the natural world. However, if biophilia is an instinct—or, more precisely, "a complex set of learning rules" philogenetically adaptive (Wilson 1993)—it can remain silent for the entire life of an individual unless it is appropriately stimulated. Human functions that regulate our relationship with the natural world can "persist from generation to generation, atrophied and fitfully manifested in artificial new environments into which technology has catapulted humanity" (Wilson 1993).

Thus, how can biophilia be expressed in modern society? Howard Gardner reckons that naturalist intelligence, which, in his theory of multiple intelligences is the eighth form of human intelligence, is displayed in those people who take care of, and subtly interact with, the living creatures, such as deep ecologists and naturalists, but:

> even apparently remote capacities—such as recognizing automobiles from the sound of the engines, or detecting novel patterns in a scientific laboratory, or discerning artistic styles—may exploit mechanisms that originally evolved because of their efficacy in distinguishing between, say, toxic and nontoxic ivies, snakes or berries. Thus, it is possible that the pattern-recognizing talents of artists, poet, social scientists, and natural scientists are all built on the fundamental perceptual skills of naturalist intelligence (Gardner 1999, 50).

Hence, in the industrial world, biophilia appears to take the resemblance of an exaptation, a characteristic that evolved for a particular purpose but that reveals itself useful for a different one. As educators, we feel we have the duty to recover biophilia as an instinct; that is, "the innate tendency to focus upon life and lifelike forms, and in some instances to affiliate with them emotionally" (Wilson 2002, 134) in its original form of evolutionary adaptation: the development of a deep and healthy relationship with nature. Undoubtedly, we have the need to develop an educational project that can stimulate biophilia, but more in general we need a framework for environmental education that starts from biophilia to nurture in the child, the adolescent and the adult the awareness that a profound relationship with nature is necessary for a harmonious development of one's own personality. If this is the conceptual framework of our research, we can begin to ask ourselves more specific questions, such as: What meaning and what value can we give to the different experiences— either direct, indirect or symbolic ones—with nature that children have during their childhood? Is it true that during childhood children develop profound connections with the natural world that are then suppressed during adolescence? How can educators help adolescents to recover such connections with nature without undermining their need for social interaction? How can adults be helped to recover their own feeling of belonging to the natural world?

The formulation of such questions is not conducted by chance: it follows the order of ontological development of the child toward adult life and the related continuous adjustment of the needs and necessities in the different life phases from childhood to adulthood. Such differentiation should find correspondence in the educational experiences at the different levels. It is not difficult to notice that current school education keeps prioritizing cognitive aspects at the expense of more profound levels of connection with the natural world.

Thus it appears important to develop a research area that could be defined *affective ecology* and could be an area of ecology that deals with the origin, growth and maturation of genetically determined and evolutionary adaptive, affective relationships between human beings and other living organisms. Affective ecology is an integral part of the process of affective appraisal of environments, which is the conferral of affective qualities to the environment that is strictly connected to environmental preference. Affective appraisal is also one of the components of environmental schemata (along with the cognitive, behavioral, affective and evaluative components) that are the knowledge structures that organize environmental information deriving from perception and that guide behavior (Berto 2002), hence contributing to the process of knowledge of the environment. Profound parts of this affective process of knowledge of the environment are what Louise Chawla (2002) called "magic spots of time," when children experience an insightful connection of tenderness and love that they have with living creatures and they become aware of the need for the care and attention that they require. However, the possibilities of realizing magic spots of time when immersed in nature are few and far for school-age urban children. In addition, such children are prevented from enjoying the process of regeneration of direct attention created by the fascination with nature, which occurs by simply passing time immersed in it (Kaplan 1995). Then, it so happens that the direct attention of children who live in urban areas is quickly exhausted, as it is continuously attracted by enormous quantities of sensory, emotional and cognitive stimuli. These are children who are deprived of their need for magical moments and the restoration that can be found in nature. In such situations of limited contact with nature, Stephen Kaplan proposes the practice of meditation as a means to reinforce the restorative power of direct attention. Kaplan suggests that people

> with little meditation training attempting to meditate in an environment arranged to have only modest restorative properties can experience more recovery of directed attention capacity than either the same person in the same environment who is not attempting to meditate or the same person trying to meditate in an environment that offers fewer restorative properties (Kaplan 2001, 500).

We wanted to put into practice Kaplan's hypothesis, using the practice of mindfulness as a tool for the recovery of direct attention, as well as the recovery of the dimension of direct contact with nature, starting with awareness of one's own senses (Kabat-Zinn 2005). In its essential form, the practice of mindfulness is a practice of active silence, which offers the opportunity to experience moments of suspension from the multiple aural and visual stimuli and to enter into a relationship with one's own internal space. On such premises, we developed an experimental educational activity of *active silence training* (AST) addressed to primary children, with the aim of stimulating biophilia through the recovery of attention (Barbiero et al. 2007). One hundred twenty children from a primary school in the town of Aosta, Italy, took part in a series of experimental activities that were part of a study aimed at evaluating the efficacy of AST in the process of direct attention recovery. AST makes use of silent observation as a tool for knowledge of oneself and of one's own body, and play as a moment of fascination involving involuntary attention (James 1892), allowing voluntary attention to rest and regenerate (Kaplan 1995). The experimental protocol set out to measure some baseline physiological parameters and a series of tests of direct attention before, during and after AST. The results appear to support Kaplan's hypothesis: as compared to the control group, children in the experimental group were significantly quicker in the direct attention tests (Barbiero 2007). If the practice of active silence was proved to be able to recover children's attention capacity, it could constitute the basis of an educational program aimed at revealing children's biophilia and a new and original way to develop their naturalist intelligence (Gardner 1999).

NOTES

1. Ministerial decree of May 26, 1998. Published in the *Gazzetta Ufficiale*, July 3, 1998, *n.153–Annex B.*
2. This is translated from the Italian *laboratorio,* which in contexts outside disciplinary science might be better translated as "workshop." The word *laboratory* has been left throughout.

BIBLIOGRAPHY

AA.VV. 1998. *Dall'escursionismo all'educazione ambientale. Un'esperienza sul campo. Atti del 17° Corso-Seminario Regionale per Operatori TAM,* Aosta, Italy: Club Alpino Italiano e Regione Valle d'Aosta.

Alfieri, F., M. Arcà, and P. Guidoni. 1995. *Il senso di fare scienze.* Torino, Italy: Bollati Boringhieri.

Ammaniti, M., and D. Stern. 1991. *Rappresentazioni e narrazioni.* Roma: Laterza.

Arcà, M. 2005. *Il corpo umano.* Roma: Carocci Faber.

Barbiero G., R. Berto, D. Freire, M. Ferrado, and E. Camino. (2007) Svelare la biofilia nei bambini attraverso l'active silence training: un approccio sperimentale. *Culture della Sostenibilità* 2: 99–109.

Battistini, G., A. Bezzi, B. Massa, and G.M. Pedemonte. 1982. Il laboratorio sul terreno nell'insegnamento della geologia: premesse metodologiche ed attuazione didattica, *Quaderni CIDI* 6(12): 38–71.

Battistini, G., A. Bezzi, B. Massa, and G.M. Pedemonte. 1983. Le Scienze della Terra nella Scuola Secondaria Superiore. Parte seconda: valenze educative ed implicazioni didattiche. *Didattica delle Scienze* 104: 39–43.

Berto, R. 2002. Codifica e caratteristiche del paesaggio che ne influenzano il ricordo. In *La narrazione del paesaggio,* ed. C. Gallo Barbisio, L. Lettini, and D. Maffei, 186–192. Torino, Italy: Tirrenia Stampatori.

———. 2005. Exposure to restorative environments helps restore attentional capacity. *Journal of Environmental Psychology* 25: 249–59.

Bertolino, F., A. Perazzone, and M. Bertone. 2007. L'offerta formativa in educazione ambientale nelle università italiane. Il valore di un luogo comune di sconfinamento. In *Formazione e sostenibilità,* ed. W. Fornasa and M. Salomone, 103–133. Milano: Franco Angeli.

Bertolino, F., M. Messina, A. Perazzone, and M. Salomone. 2007a. L'educazione ambientale nelle università italiane: attori, modelli, contenuti, ricerche. *Culture della sostenibilità* 1: 79–116.

———. 2007b. L'educazione ambientale nelle università italiane: attori, modelli, contenuti, ricerche. Parte II—Reti relazionali e mappe concettuali dell'educazione ambientale nell'università italiana. *Culture della sostenibilità* 2: 61–86.

Bogina, M.A., and B.R. Roberts. 2005. The use of Haiku and portfolio entry to teach the change of seasons. *Journal of Geoscience Education* 53 (5): 559–62.

Borgarello, G., and E. Trusel. 1991. *Educazione Ambientale: la proposta di Pracatinat, Quaderno n. 1.* Torino, Italy: Regione Piemonte.

Borgarello, G., A. Chiesa, and C. Galetto. 1997. *Educazione e società sostenibile. Itinerari didattici per la scuola media superiore. Pracatinat, Quaderno n. 4.* Torino, Italy: Regione Piemonte.

Cadzow, J. 2004. The bubble-wrap generation, *Sydney Morning Herald, Good Weekend,* January 17, 18–22.

Camino, E., and G. Barbiero. 2005. Connessioni, reti da svelare, trame da tessere per un cammino verso la sostenibilità. In *Per una ecologia dell'educazione ambientale* ed. E. Falchetti and S. Caravita, 101–112.Torino, Italy: Edizioni Scholé Futuro.

Carbaugh, D. 1996. *Situating selves. The communication of social identities in american scenes.* Albany: SUNY Press.

Carvalho, G.S., R. Silva, N. Lima, E. Coquet, and P. Clément. 2004. Portuguese primary school children's conceptions about digestion: identification of learning obstacles. *International Journal of Science Education* 26 (9): 1111–30.

Chawla, L. 2002. Spots of Time. Manifold Ways of Being in Nature. In *Children and Nature* eds. P.H. Kahn and S.R. Kellert, 199-225. MIT Press.*Growing up in an urbanizing world.* London: Earthscan Publications Ltd.

Cini, M. 1994. *Il paradiso perduto. Dall'universo delle leggi naturali al mondo dei processi evolutivi.* Milano: Feltrinelli.

Clément, P. 2001. Epistemological, didactical and psycological obstacles: the example of digestion/excretion. In *Proceeding of ESERA Conference: Science Education in the Knowledge Based Society*, Vol. 1, ed.D. Psilos et al., 347–9. Thessaloniki, Greece: Aristotle Univ.

Colombies, J.L. 1997. Metodologie e politiche per l'educazione ambientale in Europa. *École* 52: 38–40.

Cornell, J.B. 1998. *Sharing nature with children.* Nevada City, CA: Dawn Publications.

Ferrarotti W. 2005. Rapporto diretto con l'ambiente: condizione di significatività e validità della conoscenza. In *Atti Scuola Ambiente & Parchi*, ed. Gruppo di Ricerca in Didattica delle Scienze Naturali. Torino, Italy: Lit. Geda.

Ferrero, E., and D. Gimigliano. 2007. *Le concezioni spontanee nelle Scienze della Terra.* Asti, Italy: Regione Piemonte e Comune di Refrancore.

Ferrero, E., and M. Tonon. 2002. The field trip: an environmental workshop in an Earth Science Introductory Course for student teachers. *26th ATEE Annual Conference, Stockholm Institute of Education, RDC 2, The Training of Science Teachers.* Available online at www.lhe.se/atee/proceedings/Ferrero-Tonon_RDC_2.doc

Ferrero, E., A. Provera, and M. Tonon. 2004. *Le Scienze della Terra: la scoperta dell'ambiente fisico.* Torino, Italy: Cortina Ed.

Forni, E. 2002. *La città di Batman. Bambini, conflitti, sicurezza urbana.* Torino, Italy: Bollati Boringhieri.

Gallo Barbisio, C. 2001. *Psicologia del paesaggio: l'arte della cura e il paesaggio.* Torino, Italy: Tirrenia Stampatori.

Gallo Barbisio, C. 2002. *La narrazione del paesaggio.* Torino, Italy: Tirrenia Stampatori.

Galtung, J. 1969. Violence, peace and peace research. *Journal of Peace Research* 6 (3): 167–91.

Garcia de la Torre, E. 1994. Metodologìa y secuenciacìon de las actividades didàcticas de geologìa de campo. *Enseñanza de las Ciencias de la Tierra* 2: 340–53.

Gardner, H. 1999. *Intelligence reframed.* New York: Basic Books.

Giordan, A., and G. De Vecchi. 1990. *L'enseignement scientifique: comment faire pour que «ça marche?* Nice, France: Z'éditions.

Jaén, M. 2000. Como podemos utilizar en Geología el plateamiento y resolutión de problemas? *Enseñanza de las Ciencias de la Tierra* 8 (1): 69–74.

James, W. 1892. *Psychology: The briefer course.* New York: Holt.

Kabat-Zinn, J. .2005. *Coming to our senses.* New York: Hyperion.

Kahn, P.H. 1999. *The human relationship with nature.* Cambridge, MA: MIT Press.

Kaplan, S. 1995. The restorative effects of nature: Toward an integrative framework. *Journal of Environmental Psychology* 15: 169–82.

———. 2001. Meditation, restoration and the management of mental fatigue. *Environment and Behaviour* 33 (4): 480–506.

Kellert, S.R. 1997. *Kinship to mastery.* Washington, DC: Island Press.

Kellert, S.R. and E.O. Wilson, eds. *The biophilia hypothesis.* Washington, DC: Island Press.

Longo, C. 1998. *Didattica della biologia.* Scandicci, Italy: La Nuova Italia.

Lorenzo, R. 2000. *In città ci abito anch'io.* Perugina, Italy: Edizioni Guerra.

Louv, R. 2005. *Last child in the woods: Saving our children from nature-deficit disorder.* New York: Algonquin Books.

Luzzatto, G. 2005. Formazione iniziale degli insegnanti. In *Voci della scuola 2006,* ed. G. Cerini and M. Spinosi. Napoli, Italy: Tecnodid Editrice.

Malone, K. 2007. The bubble-wrap generation: Children growing up in walled gardens. *Environmental Education Research* 13 (4): 513–27.

Massa, B., and G.M. Pedemonte. 1998. Une recherche sur les obstacles cognitifs en géologie: quelles implications sur la formation des enseignants? In *Formation à la médiation et à l'enseignement—enjeux, pratiques, acteurs,* ed. A. Giordan, J.L. Martinand, and D. Raichvarg, 403–408. Actes JIES XX.

Maturana, H., and F. Varala. 1984. *El árbol del conocimiento: las bases biológicas del enten-dimiento humano.* Santiago, Chile: OEA.

Ministero dell'Università e della Ricerca Scientifica Tecnologica. 1998. *Decreto 26 maggio: Criteri generali per la disciplina da parte delle università degli ordinamenti dei corsi di laurea in Scienze della formazione primaria e delle Scuole di specializzazione all'insegnamento secondario.* In G.U. n. 153 del 3 luglio 1998.

Moore, R.C. 1986. *Childhood's domain. Play and place in child development.* London: Crom Helm.

Morcillo, J.G., M. Rodrigo, J.d.D. Centeno, and M. Compiani. 1998. Caracterización de las prácticas de campo: justificación y primeros resultados de una encuesta al profesorado. *Enseñanza de las Ciencias de la Tierra* 6: 242–50.

Mortari, L. 2001. *Per una pedagogia ecologica. Prospettive teoriche e ricerche empiriche sull'educazione ambientale.* Milano: Franco Angeli.

Novak, J. 1998. *Learning, creating, and using knowledge: Concept maps as facilitative tools in schools and corporations.* Mahwah, NJ: Lawrence Erlbaum Associates.

Perazzone, A. 2004. Verticale si ma come la tela del ragno. *Atti del XIII Convegno Nazionale ANISN—Una visione del mondo: Cultura, Natura, Comunicazione nell'insegnamento delle Scienze Naturali,* Bollettino dell'ANISN 13 (num. spec): 125–128.

Perazzone, A., F. Bertolino, S. Ghilardi, and E. Camino. 2002. University training for future primary school teachers; environmental education is introduced, at last!, In *26th ATEE Annual Conference, Stockholm Institute of Education, RDC 17.* Available online athttp://staff.um.edu.mt/ppac1/RDC17/stockholm2.pdf

Prensky, M. 2006. *Don't bother me mom—I'm learning!* St. Paul, MN: Paragon House Publishers.

Psarros, N., and H Stavridou. 2001. The adventure of food inside the human body: Primary students' conceptions about the structure and the function of the Human digestive system. In *Proceeding of ESERA conference: Science education in the knowledge based society,* Vol. 2, ed. D. Psilos et al., 745–7. Thessaloniki, Greece: Aristotle Univ. .

Reiss, M. J., S.D. Tunnicliffe, A. Moller Anderson, A. Bartoszeck, G.S. Carvalho, S. Chen, R. Jarman, et al. 2002. An international study of young peoples' drawings of what is inside themselves. *Journal of Biological Education* 36 (2): 58–64.

Sterling, S. 2001. *Sustainable education—Re-visioning learning and change.* Schumacher Briefing no. 6. Dartington: Schumacher Society/Green Books.

Thomashow, M. 1996. *Ecological identity. Becoming a reflective environmentalist*. Cambridge, MA: MIT Press.

Tonon, M. 2004. Le escursioni didattiche: esperienze qualificate e integranti. In *Atti del XIII Convegno Nazionale ANISN—Una visione del mondo: Cultura, Natura, Comunicazione nell'insegnamento delle Scienze Naturali*, Bollettino ANISN 13 (num. spec): 129–35.

Tonon, M., A. Perazzone, and A. Provera. 2006. Educare attraverso la valorizzazione dell'ambiente naturale: un'esperienza formativa nel contesto alpino del cuneese. In *Cultura locale e formazione*, ed. A. Rossebastiano, 155–80. Torino, Italy: Il Segnalibro.

Tonon, M., E. Ferrero, and A. Provera. 2005. Esperienze di fruizione didattica e di valorizzazione di alcuni affioramenti fossiliferi del Monferrato. *Giornate di Paleontologia 2003—Alessandria, 22–25 maggio, Rendiconti della Società Paleontologica Italiana 2*, ed. D. Violanti, 229–41.

Tonucci, F. 2002. *La città dei bambini*. Bari: Laterza.

Vasio, C. 1999. Se fossi il re di un'isola deserta—Haiku. Roma: Empirìa Ed.

Volpi, R. 2004. *Liberiamo i bambini*. Roma: Donzelli.

Wilson, E.O. 1984. *Biophilia*. Cambridge, MA: Harvard Univ. Press.

———. 1993. Biophilia and the conservation ethic. In *The biophilia hypothesis*, ed. S.R. Kellert and E.O. Wilson. Washington, DC: Island Press.

———. 2002. *The future of life*. New York: Alfred A. Knopf.

8 Role-Play as a Tool for Learning and Participation in a Post-Normal Science Framework

Laura Colucci-Gray

EDITORS' INTRODUCTORY NOTES

Building on the educational premises illustrated in the previous chapters, we now turn to the specific learning and teaching methodology of role-play, and how this can be used to create an educational context for dealing with the complex, transdisciplinary problems of sustainability. As reported earlier, sustainability education poses challenges to the traditional context of science learning. This chapter examines the proposition of a role-play dealing with a complex, global issue to engage students in developing competences that are commensurate with the transdisciplinary and dialogical context of a society in post-normal conditions. In the final part, an illustration of a recent experience of role-playing in a secondary school classroom will focus on the process and outcomes of students' engagement in the role-play. This chapter has many purposes: to establish the theoretical foundations for using role-play in a framework of sustainability; to illustrate how a role-play can be constructed and used for this purpose and to exemplify the learning processes that are enacted. This will form the basis for a reflection on the changing nature of knowledge production in the classroom and the wider implications that this can have for education and research at the interface with science and society.

SCIENTIFIC LITERACY AND THE FRAMEWORK OF SUSTAINABILITY SCIENCE

The perspective of sustainability science (Gallopin 2004) is proposed here as a conceptual framework for developing a type of knowledge that is conscious of itself and collaborative:

> Sustainability science needs to explore, apply and teach a vast array of instruments and methods, rather than focusing only on a few. It will be receptive to new ideas and visibly more holistic of current science. By embracing uncertainty and incorporating the qualitative aspects it will

bring to an enormous broadening of the universe of possible solutions (and questions). (384)

Hence the framework of sustainability science, which acknowledges complexity, interdisciplinarity, partiality of views, precaution, humility, equity and reversibility of impacts, has been introduced as a framework for presenting the educational experiences set out in Part 2 of this book. Now, this chapter focuses on a particular learning and teaching strategy, that of role-play, which is proposed as a tool for introducing students to approaching problems in post-normal science conditions, motivating and supporting the development of specific competences. As reported in the introduction to this book, the methodological choices and value-frameworks become important. In particular, the qualitative proposition of nonviolence, which is rooted in relationships of equivalence, active listening, respect and dialog, is considered here as the founding element for developing an inclusive framework of dialog and co-existence within the boundaries of the Earth.

What follows is a review of the epistemological and pedagogical features of role-play within the outlined framework and an analysis of possible implementations in the classroom. The purpose is that of reflecting on some of the choices that could be made from a teaching point of view for structuring the learning environment in a way that is conducive to addressing complex problems. First, I will start with an analysis of the topic of the role-play: this is presented as an example of a complex problem facing society, with implications at both the global and local levels. The case will resemble the issue presented in Chapter 3, and its analysis will illustrate some of the "learning signposts" that will populate the students' learning environment through the role-play. From this initial reflection, I will then move onto the pedagogical aspects, with a particular focus on the features and purposes of role-play simulations and their place in classroom learning.

THE COMPLEX AND CONTROVERSIAL SOCIOENVIRONMENTAL ISSUE OF INTENSIVE PRAWN FARMING

Many issues in the past twenty to thirty years have been characterized not only by groups of experts with divergent views, but also by clashes, sometimes of a violent nature, between different social groups. Such clashes could be associated with competition over the use of resources, or they could have arisen from the experience of hardship derived from the depletion of important resources and environmental services. Among the most common and notable cases, we can cite the construction of dams, such as the one on the Narmada River in India and, more recently, intensive cultivation for biofuel production. When we devised the role-play described here, intensive farming of prawns in the southern part of the planet was in rapid expansion. Supported by international institutions, such as the Food and Agriculture Organization, the International Monetary Fund and the World Bank, intensive farming of prawns was aimed at boosting the economy of developing countries and improving the protein intake in the diet of the rural and fishing communities of the areas involved. In a few years, this program spread along the coasts of Southern Asia, South America and Central Africa to emerge as one of the most dominant food-producing and revenue-generating sectors (Bunting 2007). Prawns produced from aquaculture have been regularly exported from the southern producers to the western consumers in the United States, Europe and Japan. Their production, however, has been the cause of widespread damage to the coastal ecosystems, weakening the subsistence economy of local populations and generating inequities and violent conflicts between local communities and farm owners.

The controversy over prawn farming has similar characteristics to the others cited before: it has acquired a global dimension, and in so doing, it has boosted transformations of large extensions of natural ecosystems, with displacement and use of great quantities of materials and energy, thanks to a concentration and joint deployment of financial, economical and political powers. In most situations, the activities were promoted and encouraged by the scientific communities, experts acting from the academy or from international institutions (Food and Agriculture Organization, United Nations etc.). But often what seemed a good idea according to the results obtained within the boundaries of research labs can lead to unexpected consequences, and give rise to multiple feedbacks, extending well beyond the original spatial and temporal context. The complexity of the real world, alongside the complexity of the multiple views held by the many actors involved (local communities, politicians, traders, institutions and the consequences on humans, as well as on other living beings) led to controversies in which the multiplication of voices and interpretations made it increasingly difficult not only to find the "right" solution, but also to put decision-making processes in place that would involve all stakeholders. Even among the scientists, there have been different schools of thought, which, over the years, have held contrasting positions[1].

From a social point of view, particularly relevant was the nonviolent movement in the southern Indian state of Tamil Nadu that was organized by local villagers against the growing shrimp industry, in line with the Gandhian tradition of Satyagraha (Rigby 1997). The movement raised awareness of the issues of inequity associated with the growing shrimp industry, and on December 11, 1996, following the opportunity to be heard by the Indian Supreme Court, an order was issued in their favor against the local farming industries.

However, intensive prawn farming in India and around the world has continued to expand. The problem is therefore still open, trans-scientific, involving knowledge as well as different perceptions of the environment held by different stakeholders, and it cannot be dealt with only at the local level. Here, we can identify the complexity of science–society interactions and decision-making processes, the analysis of which can provide important pointers for setting up the educational framework of role-play. We begin such an analysis in the next section.

Partiality of Perspectives, Ecological Complexity, Uncertainty and the Problem of Consensus in the Prawn Farming Issue

The topic of the role-play proposed here features many aspects of the science–society debate illustrated earlier, as well as in Part 1 of this book. Similar to the controversy described in Chapter 3, prawn farming has been characterized by cycles of controversy about the environmental impact of the ponds, as a series of scientific studies conducted by different agencies contradicted one another in succession (Rigby 1997). An important component of the controversy over the facts is constituted by the inherent complexity of the coastal ecosystem, which was underplayed, somehow overlooked when prawn farming was originally launched. In the local ecosystems of southern India, and (as it was experienced) in all ecosystems characterized by the mangrove green belt at a planetary level, the mangroves play a complex ecological role. They provide a "nursery" environment for many species of birds and fish, including prawns, and they protect the inner land from the sea. The clear cutting of the mangrove trees to free up surface for the construction of prawn farming triggered a number of consequences: 1) the loss of mechanical protection of the coast made the internal areas more vulnerable to the destructive impact of the sea—on a number of occasions, monsoons, floods and, a few years ago, the tsunami, swept away the remaining defenses, leaving the agricultural land unprotected; 2) the replacement of the natural mangrove ecosystem with the artificial ecosystem of the pond for intensive farming, featuring only a few selected varieties of animals, brought a reduction in biodiversity. In addition, biodiversity had further decreased because the mangroves were used as "nurseries" by many species (fish, birds, invertebrates) that migrate in their adult life toward other environments. So, food chains and ecological webs have been disrupted far beyond the local boundaries. The complexity of such interactions was little appreciated or even recognized when industrial aquaculture was first promoted and subsidized, mainly owing to the specialized, reductionist approach to the issue. Scientists were engaged in selecting larger size strains of prawns in view of more productive yields and were not keen to explore the implications for the larger context. In recent years, ecological complexity has been gradually deemed as crucially important, together with the acknowledgment of the dimension of uncertainty and ignorance in our knowledge of ecological systems.

In addition, it is also true that scientists (as lay people) are embedded in a sociopolitical context, use a particular language and investigate problems not only from different disciplinary perspectives, but also according to different priorities, stemming from a personal vision of the world—from cultural influences or from the demands of those who have requested the research. Hence, the nature of scientific evidence, as well as the social context in which such evidence is produced, is bound up with the socially accepted values, norms and consensus of a particular group. This brings us to consider another aspect of controversy, which is the broader scenario of competitive action (Dimitrov 2003): when different kinds of evidence are confronted, those who are in a position of power prevail, and this is not only the fault of a particular social system that may favor collusion over transparency. It is actually a feature of our own way—as social beings—of selecting information and making value judgements. As reported by Chambers (1997): "People only see what they are prepared to see. [. . .] We create what we want to see; and the more powerful we are, the more we do this, and the more it is done for us" (100). In this context, a power position can come from the persuasiveness of an expert's interpretation, depending on their ideas and whether they reflect larger consensus in the community, the abilities of the individual or the reputation of the belonging group (Moscovici and Doise 1991). From this, it follows that the use and production of scientific evidence is far from being neutral and objective: it can generate acceptance, dissent, but it can even be completely disregarded. An open discussion based on evidence would thus be an oxymoron, as it would necessarily be limited by the nature of the context in which the discussion takes place. Even material resources, such as the means available for producing, communicating and sharing the evidence, would affect the shape of the discussion and the nature of consensus. From these considerations, we can derive three main lessons, which can provide the assumptions underpinning an activity of role-play:

- Problems emerging with prawn farming cannot be "solved" by any single answer or source of evidence. It is an interdisciplinary issue, and each disciplinary perspective can only grasp some aspects of reality, which are analyzed through a specific approach and have to be integrated to produce a significant interpretation.
- The real world is complex and the views of each disciplinary perspective are not only insufficient because they are partial, but because the very phenomena that we are studying are in the process of becoming: new configurations of the system are expected to take form all the time, according to the natural evolutionary trend. Hence, knowledge is not only partial, but also contingent, linked to space, time and context. A science that is mindful of such aspects will embrace and account for the ignorance of both the present system and its future configuration.
- The different points of view expressed by the various experts depend on many aspects and they are never conclusive. When involved in discussion and decision-making processes, the experts' points of view

can prevail on the viewpoints of other witnesses, lay persons and local communities because of the greater legitimacy they receive from the institutions and society at large: legitimacy that gives power in decision-making processes. For example, in the controversy of shrimp farming, the voices of the local people reporting the damage to the local environment had not been considered until recently, when the collapse of tropical ecosystems due to intensive aquaculture was officially described in scientific reports.

- Knowledge of the system is not cumulative, but dynamic and changing, requiring to be reframed and reflected in negotiation with others.

Such points have important implications for learning and education. The issue of prawn farming illustrates the limitations of our knowledge intended as a definite description of reality, and it has emphasized, from an epistemological point of view, the need for deeper and fluid engagements between the observer and reality (other people, living beings, natural contexts) that is being experienced:

> [But] now we know more about what is not knowable using the standard methods of established disciplines. When so much is so unknowable and so unpredictable, it seems sensible to seek solutions through methodological pluralism, through flexible and continuous learning and adaptation (Chambers 1997, 193).

Language and awareness of its uses can be one of the vehicles for accessing such methodological pluralism and for researching the connections with the contingent reality we inhabit. We will look at this in the following section.

Language, Science and Educational Context

As reported in Chapter 4, language and metaphors are tools for creating reality. We express ourselves by means of language—that is, verbal and nonverbal, words and actions—that continuously express an intimate interconnection between facts and values and an inevitably partial perspective. Such partiality is not only located at the level of the individual, but—as we have seen previously—it is embedded in the very existence of different societal groups, which are organized around particular discourses and norms in society. Discourses can be powerful elements in strategic interactions, as they are tied to history and culture, and they may be institutionalized, remaining dominant in the attitudes, metaphors and images that people use when referring to social and cultural practices. Gee (1999) defined capital "D" Discourse as comprising the "socially accepted associations among ways of using language, of thinking, valuing and acting, and interacting, in the right places and at the right times" (17). Hence, in complex situations, which are characterized by people holding not only a variety of points of view, but also different, deeply rooted cultures, worldviews and languages, there can be a multiplicity of means for expression and interaction, and some discourses can be very powerful in influencing people's perceptions and choices. They can determine what kind of scientific knowledge is being used, privilege certain models for looking at reality over others or what knowledge and voices should be sought and heard.

In the case of prawn farming, for example, the discourse of unlimited growth called for research into the making of disease-resistant species that could boost and support intensive farming production. In contrast, the discourse of ecological interdependence called for studies in ecosystem dynamics. In addition, the issue of prawn farming featured elements of conflicts between groups that were not equally legitimized. So, for example, the discourses of community-based living were overlooked and materially hindered by the promises of economic development supported by scientific and technological means at a large scale. Besides, the voices of the farmers denouncing the negative impacts of the farms were further delegitimized and discarded because of a pecking order in views of knowledge that assigned less importance to local and anecdotal experience considered of little scientific value.

Hence, such discussions point to the question of how to deal with sustainability issues and the ephemeral nature of consensus. Folke et al. (1998) pointed to the variety of contrasting modes of thinking in the ecological, economic and social sciences. A first model is based on an image of the ecosystems as stable, whereby a change would be temporary and controllable before reverting to a previous state, and essentially made of resources that can be extracted for human use. Another model starts from the notion of *humans-in-nature*, or a socioecological systems approach (Berkes and Folke 1998), in which the delineation between social and natural systems is artificial and arbitrary, and as we have seen in Chapters 4 and 6, the boundaries between them can be perceived as real, as well as shifting. In addition, models of research valuing large-scale, quantitative and generalizable studies would sometime compete over other models of inquiry based on qualitative aspects and first-person experiences.

Within the framework of sustainability science and nonviolence, an opportunity arises for multiplying, balancing and legitimizing the different views in such a way that: "acknowledging the partiality of my perspective (in the double sense of preferred as incomplete) encourages me to place my own views in question by considering the partial basis on which I hold them" (Primavesi 2000, 10). Different communication processes can be associated with a scenario of multiple perspectives. One is based on discussion that, from the Latin *discussio*, meaning "to shake and break something apart." Argumentation can be a powerful feature of discussion, and some authors have argued for this skill to be taught for the purpose of learning the analytical processes of reasoning and building knowledge in science (Osborne 2003). The other perspective is based on dialog, from the ancient Greek *dialogos*, meaning "a verbal exchange between two or more people involved in conversation." Like the Socratic dialogs, a dialogical conversation involves people who talk together to investigate a subject in depth with an open spirit of inquiry. Within a dialogic interaction, the skills of persuasion (Billig 1996) can be used to invite the other to see the world from another point of view, yet without coercing him or her into any preferred way of thinking and talking. As reported by Kabat-Zin (2005): "The quality of the relational space is crucial for the purpose of opening new spaces and new manifestations" (449).

In the realm of education, this can be taken as a suggestion for establishing interactive processes in which students can tackle the interconnections between modes of thinking, knowing and talking in socioenvironmental issues, and where they can practice with laying down new worlds of sustainable co-existence. In the following section, I will go into details as to how role-playing can be used to support this kind of educational process.

THE EDUCATIONAL PREMISES OF THE ROLE-PLAY METHODOLOGY

In line with the educational approaches described in previous chapters, the analysis of the issue of prawn farming highlights the emerging awareness of the need for a dialogical, multimodal form of hybrid knowledge, arising from the interconnections and exchanges between multiple perspectives. However, introducing such a view in the science classroom throws up multiple questions; for example, on the nature of learning and how this can be re-conceptualized beyond the transmission of concepts, and the nature of the relationship that can be considered appropriate between teacher and students. Other points that are immediately related to these refer to the idea of educational research and the role of the researcher. Is it more appropriate to observe the processes from an external point of view, choose a few significant variables, collect data that can be quantified and allow for an objective assessment of what is being learned, or is it more suitable to initiate a process of action–research, with the researcher entering into dialog with the students and participating with them in a process of learning on a multiplicity of levels (cognitive, relational, linguistic)? The mode of doing and reporting research will also change from a search for truth and objectivity to a truthful representation of the learning process[2]. Hence the epistemological assumptions underpinning the use of any particular methodology—including role-plays—the modalities in which the research is being carried out and the choice of what is considered significant are all aspects that concur to shed light on the complex and problematic features of the context of learning.

For the purpose of framing role-play within the context of post-normal science, it might be useful to summarize some aspects derived from previous reflections:

- knowledge is dynamic, partial and in continuous transformation;
- knowledge is contingent: what we know and what we are able to know is closely interwoven with how we know and the nature of the context;
- language, like knowledge, is more than simply passing information, but linguistic interaction occurs within a context of values, social practices and norms that shape the nature of the meanings that are being produced (Varela et al. 1993);
- awareness of the different representations of the world is a step toward building awareness of complexity and a way of knowing which is more respectful of other people and other living beings;
- transdisciplinary links can be made by means of dialogical approaches that lead to self-expression, sharing, listening and juxtaposing scenarios and contexts that may be apparently unrelated;
- the classroom is a space for making new connections and new discoveries that are collectively and personally relevant.

Hence, for the purpose of elaborating an educational reflection on role-play within the context of post-normal science, attention will be given to the development of competences arising from activities that encourage epistemic development (i.e., definition and framing of problems and concepts), redefinition of values and assumptions about the role of science and the experts and co-construction of realities. We now turn to the pedagogical aspects of role-play.

Pedagogical Features of Role-Play

Role-plays are associated with a specific category of active learning techniques, known as gaming and simulations. Ellington (2000) distinguishes among three main exercises: simulations, games and case studies, although there are many areas of overlap and hybridization between them and such tools can be flexibly combined to create different contexts for learning. For the purpose of this chapter, I refer specifically to particular role-plays that contain elements of simulation, dramatization and gaming. The most peculiar feature of role-plays that include a dimension of simulation is the possibility to construct structures and contexts for dynamic interaction that can be used to mirror societal situations. Simulation exercises can represent a real or realistic situation of some sort and they are usually ongoing, i.e., in progress. Reality may be, to some extent, distorted, as some elements may be made simpler, partial or be pulled out. Some simulations offer the opportunity to modify spatial and temporal scales by presenting, within a couple of hours, events that occur over years. In addition, a simulated scenario can be enriched by adding features of gaming. As defined by Adams (1973), games may have "an end, a payoff and there are explicit game rules to follow in order to reach the payoff" (3). Hence, games add elements of strategic interaction to the simulation: many games tend to have winners and losers, and for such a reason, players usually adopt competitive behaviors, although there can be moments during the game when cooperative behaviors are also encouraged.

For such features, simulations can be used to recreate contexts for participation in and experience of socioecological issues, such as prawn farming as discussed previously. So, for example, within a simulation it is possible to recreate a situation in which people who are normally scattered all over the world meet—for example, at a conference—and interact directly, and different rules of participation can influence the dynamics of interaction. Participants can take sides or they can put themselves in a position of reflection and search for the values and assumptions underpinning their actions.

In among the many possibilities of simulation that can be encountered, role-plays can stimulate multiple levels of involvement. First, a scenario can be devised and students can prepare themselves to take on the role of another person, real or simulated; by so doing, they will take a bounded perspective and will have to express the knowledge, values and emotions of this person in the simulated context. Taking on a role allows students to tap into their previous knowledge of the situation, as well as their stereotypes, discourses and perceptions. Social practices that students can witness in real life are then dramatized, and this creative effort allows for the variety of meanings, perceptions, assumptions and values to be disclosed and reflected upon in a public forum. In addition, while role-playing and dramatizing, they bring their experienced reality to bear against the often simplified and predictable reality of the classroom. In line with Barab and Roth (2006), "participation" is presented here as being directly connected with learning. More than simply exposing students to the facts of reality, the activity of participation in a role-play aims to activate students' lived experiences, meanings and identities and to engage them in understanding issues that are still open to debate, are unsolved and require new imaginations. Students' ways of learning are valued and their prior knowledge—formal and informal; personal and practical—is activated to support investigative tasks. They engage with the curriculum of everyday life, which is not formalized and codified into assessable concepts. For such reasons, some authors have described the act of role-playing as an act of temporary protection that allows students to deal with contentious issues, with dissent and with an unconventional use of knowledge (Mitchell 2000). In addition, an understanding of what often appears to be only social conflicts have an important (and hidden) component linked to an unbalanced accessibility or use of natural resources and services, would also bring students to question our ways of knowing the natural environment, the assumptions that we bring about the natural world and how these lead to actions that impact the Earth and other living beings in different ways. This may be challenging for some teachers, who find themselves in the unusual position of participant spectators, who can learn from and be affected by the process of learning, rather than controlling it or trying to provide the ultimate answer that would settle the issue. It is indeed the class as a whole that is invited to reflect on the process and the (temporary) consensus that has been achieved.

The activity unfolds through a balance of improvization and interpretation within a structured context. Hence, for such characteristics, role-play simulations are also categorized as exercises, as they require participants to develop competences for acting in accordance with the context that is being created. In some cases, students can rehearse skills of self-expression, argumentation, empathy that they may already hold, or it may be a new challenge. Afterwards, the whole community can reflect on both the process of learning as well as on what is being learned. Such learning processes can have important consequences for epistemic development. As indicated in Chapter 4, the analytical processes of thinking that tend to divide, oppose and dichotomize tend to be developed at the expenses of integrative processes of thinking. Arguably, role-play can be proposed for overcoming such an imbalance by developing awareness of the polarity that exists between dualism and nondualism, and by experimenting with dynamic situations in which participants are pushed toward the exploration of both extremes. Such explorations can take place at different levels. For example, while players take on a role, they can perceive themselves as separate from the characters or connected to them by feelings of empathy. Similarly, during decision-making processes, they may feel the need to argue and discuss; they may compare what can be understood "scientifically" with what is being expressed by other languages and representations, or they may engage with dialog and the search for common needs. Hence, in the context of the classroom, players can be introduced to a context in which different forms of knowledge are juxtaposed and examined to allow for new connections to be made between concepts or contexts that are not normally perceived as being connected. This can produce players' awareness of how society works and their own ways of living and acting.

The diagram presented in Figure 8.1. was elaborated for the purpose of bringing together the multiple dimensions of a role-playing activity as a learning experience embedded in the epistemological framework of post-normal science. The diagram was devised in collaboration with Elena Camino following a sustained experience of preparation and experimentation of a variety of role-plays on socioenvironmental issues in different educational contexts (Colucci-Gray et al. 2006). Hence the framework encompasses a multiplicity of dimensions of learning, and it can be used as a tool for understanding the learning processes occurring in different role-play situations depending on topic, age of the students and specific contexts. Starting from the inner circle, we can observe a series of nested competences:

a. Individual level: students are invited to examine their own opinion in the light of the opinion reported in the role-card, develop empathy, search for information. When they work in a small group, interpsychic aspects are also stimulated. Students have to establish relationships with other members of their own group and be able to express themselves, listen, compare and integrate the different points of view, divide tasks and give responsibilities. At this level, they practice with the first element of participation in an extended peer community: mental representations and misconceptions can emerge. Peer learning can also occur, and the need to take on a role can lead students to practice with abilities of dramatization that are unusual in a school context.

b. Collective level (interactions between groups): different ways of playing the game can require students to behave in different ways and according to different objectives. In a competitive setting they have to think strategically, argue and coordinate themselves in order to compete effectively to achieve a specific goal. In contrast, in a scenario of dialog, other dimensions, such as the ability to think about a shared future, to listen and to be receptive to other people's needs also emerge.

c. Ecological level (human actions within the natural systems): through the simulation of a real environmental issue, students can become aware of the complexity of ecological systems and the networks of interdependencies connecting natural processes and human actions. Hence, they can appreciate that there might be unforeseen consequences and feedbacks that are often unknown. At this level, it is possible to develop learning of the natural sciences with specific concepts (ecosystems and their relationships, evolutionary processes etc.), but also initiate a reflection on the idea of science and build the basis for ethical reflection.

Starting from the center, Figure 8.1. shows the competences that can be developed at each level, and it could be used by the teachers for planning an interdisciplinary activity or verifying the students' levels of learning on realms that are more articulated and complex than what is usually being considered and assessed. To proceed to the analysis of learning processes through role-playing, we can now turn our attention to the specific features of the role-play on prawn farming and the experience of a group of secondary school students' engagement with this role-play as an illustrative example. This will allow us to draw some conclusions about learning through role-playing, and particularly the role of science learning.

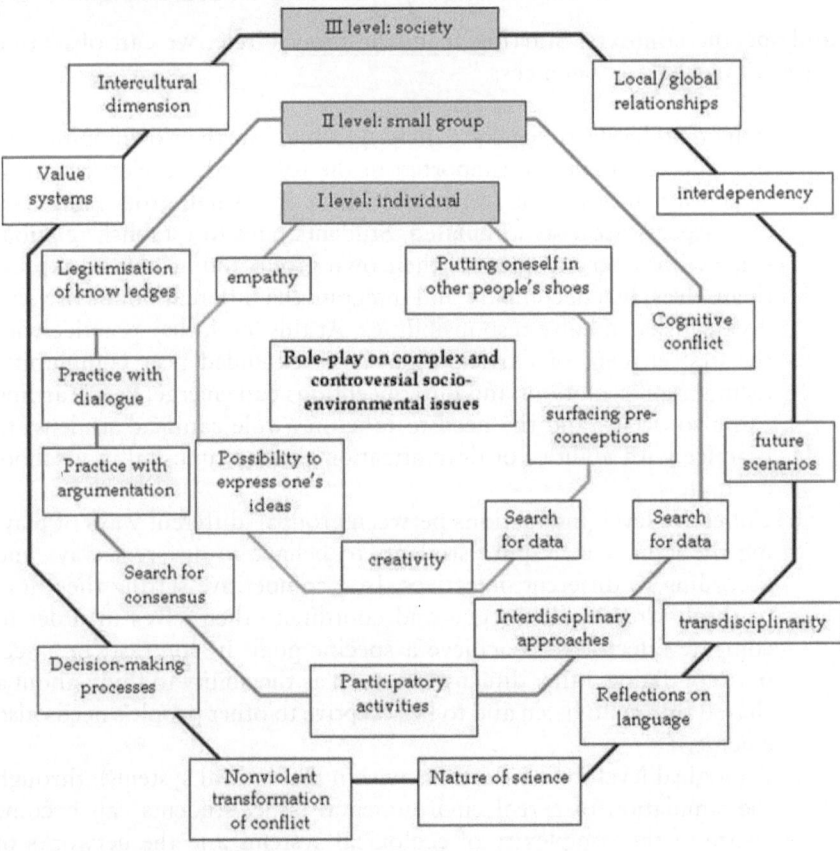

Figure 8.1   A framework for learning through role-plays on complex and controversial socioenvironmental issues. © L. Colucci-Gray and E. Camino, original work.

A Role-Play Simulation Dealing with the Issue of Intensive Prawn Farming in India

Drawing on the methodological features of role-play described previously and the diagram in Figure 8.1., the role-play on prawn farming (Colucci and Camino 2000) involves participants in the simulation of public, decision-making processes featuring different kinds of participation and interaction between stakeholders. Two settings are created. The first setting simulates the Court of Inquiry: in this context, students are members of different groups of opinions, presenting their case in front of a panel of decision makers. They are divided in two groups holding contrasting viewpoints, and each group aims to prevail over the other. In this setting, also named win-lose, aspects of the science–society interactions are represented, such as:

- different, disciplinary and fragmentary knowledge of the ecosystems;
- manipulation and selective use of evidence to defend interests;
- judgments based on expert opinion;
- use of language to argue and persuade others in order to win;
- building of aversion and bad feelings toward the other party;
- trying to win in a short-term scenario.

In the second setting, participants are engaged in a round table, according to a model of participatory democracy (Davies 2003), resembling the model presented in Chapter 2, in which the aim is that of finding a "win-win" solution through dialog. Such an approach is in tune with the situation we came across in Tamilnadu, where local movements were struggling to cope with conflict in a creative, nonviolent way (Galtung 1996). In this context, other aspects of social interaction are valued, such as:

- an integrated, transdisciplinary view of the ecosystems;
- exploration of common needs;
- search for a multiplication of legitimate views;
- use of language to express oneself and to show one's own experience in order to build common understanding;
- to distinguish the problem from the person and build empathy;
- visioning of a long-term, shared future.

Implementing the Role-Play: Illustrative Examples from a Recent Study

As indicated earlier, this role-play had been tried out in different contexts. Here, I will draw on an experience conducted in England, which involved secondary school students, aged thirteen to fourteen years old, over a period of two hours. The topic fitted well with general science topics in secondary schools: food chains and food webs, the interdependence of organisms within ecosystems, human environmental impacts and the role of science and technology in supporting human development, with the associated ethical issues. The implicit assumption that guided the intervention was that students would be able to engage in interdisciplinary learning and grapple with the dynamic, contingent and partial aspects of the nature of science. The research was organized as an empirical investigation in a natural context, in which something could be learned about the methods, the process and the context (Colucci-Gray 2007).

Activity Set-Up

The activity was designed in three different stages, from information seeking and discussion in groups, to presenting in role and dealing with conflict.

I. Preparation. The topic of the controversy was introduced and presented to the students as a real issue, open and unsolved. Photographic material was used to present the Indian coastal environment with the mangrove trees, the prawn farming installations and the type of farmed prawns, the environmental impacts and the people involved in the controversy. Group work was then introduced in the first part of the activity to support role-playing. It was felt that group work would encourage students to discuss their role, to assimilate the relevant information and acquire competence with an unfamiliar part. To this end, students were divided into friendship groups of three, and each group was given a single role-card, with the specific remit of discussing it, finding additional information and producing a persuasive argument[3]. In addition, a group of four students constituted a commission of adjudicators. Their character was summarized briefly in an individual role-card containing a set of questions, designed to guide them through their task of questioning and listening to the characters during the simulation.

II. Dramatization. This included role-enactment in the context of a Court of Inquiry. After the adjudication and the declaration of the verdict, an activity of individual reflection was planned to gather students' feelings about being in role during the simulation.

III. Dealing with conflict. Students in role were reorganized to form larger groups, with roles of opposing points of view being represented. Students were briefed about the aims and practices of conducting dialog in a conflict situation.

In the final part of the activity, students were asked to express their thoughts about the role-play simulation exercise. Additional evaluations from the students were gathered at a later stage, and all students' conversations were tape-recorded.

Characters and Role-Cards

As a common characteristic, each role-card would contain some biographical information and examples of life experiences to which the players could refer for their enactment, and a simple and deliberate language was used (Van Ments 1999). The role-play was designed to allow students to make their own interpretations, and it was the students' task to draw on the characters' profiles and to interpret their points of views. This particular role-play has been played in different ways, with either a large or a reduced number of characters. For the specific characteristics of the English study, with students playing in small groups, the number of characters was reduced to eight witnesses and four adjudicators (see Table 8.1.).

Table 8.1 Choice of Characters for the Role-Play on Prawn Farming

Generally in favor of prawn farming	Generally against of prawn farming	Commission of adjudicators
Sonja Rey (Minister for the Development of India)	Tami Sunethra (representing the movement for the land)	Robert Brown (Minister for Indian Agriculture)
Shailesh (Indian landowner)	Margherita Broecarts (Ecologist)	Priscilla Singh (Representative from Food and Agriculture Organization)
Dr. Krishna (doctor)	Jeganatthan (leader of the nonviolent movement)	Dr. Goshivah (doctor)
Paul Power (American entrepreneur)	Dharwar (head villager)	Marco Dandri (Italian nongovernmental organization volunteer)
		Anita Randrapradesh (medical researcher)
		Satish Rampal (ministry for Indian education)

As Table 8.1 shows, characters ranged from local Indian villagers and local landowners, to representatives of the Indian government and international organizations, professionals and foreign entrepreneurs. In line with the characteristics of simulations, the list of characters constitutes a fairly simplified range of representatives that were, however, selected as playing key arguments[4] in the controversy and providing different perspectives on the same issue. For example, Paul Power would represent the foreign entrepreneur, looking at India as an expanding market for business. In contrast, Jeganatthan[5] would represent the "other face" of the Indian population, namely the majority of Indian people, who live in rural areas. Jeganatthan would then describe his efforts for the redistribution of the land to the poor to increase the self-sufficiency of the families in rural villages. Finally, the adjudicators have a position of responsibility: they can be ministers, doctors or activists, and they have the task of studying the issue and finding out more about it from interviewing the characters and evaluating the evidence.

Data Analysis

Analysis of data comprised students' recorded discussions, as well as written comments and reflections. The analysis was aimed at capturing the "holistic" and qualitative aspects of students' learning. Hence, the analysis evolved "from categorising single particles of talk to the identification of storylines, made of conversational patterns and contexts" (Colucci-Gray 2007, 121). The framework of collaborative talk devised by Mercer (2000) was adapted for the purpose of combining conversation and discourse analysis (Potter 1996, 133).

The Voices of the Students: Selected Examples

Group Work (Stage 1)—Taking on Role and Epistemic Development

This excerpt shows an example of collaborative work in pairs, displayed by a series of closely connected exchanges of collaborative thinking:

> *Girl 1:* So basically we . . . yeah, we could persuade the government that they are very risky because ehm . . . /
> *Girl 2:* Well, (because it offers no protection)
> *Girl 1:* (And it makes) the land full of salt, which means that we can't grow anything even after they have gone,
> *Girl 2:* okay. So it is basically what that . . . ehm is? Is what is called cause and effect, they are not thinking in the long run basically
> *Girl 1:* Yes because it means that nothing will grow there so eventually their whole land will get covered in it.
> *Girl 2:* I think we need to make notes really (pause) *noise in the tape-recorder*
> *Girl 1:* Okay also, also, also

In this form of talk we can recognize the beginning of role-taking through the use of the shared first person ("we"). Additionally, we can observe the influence of role-taking and cooperative talk on the girls' conceptual understanding of the issue. For example, the girls juxtaposed the perspective of a local person, i.e., a farmer: "Girl 1: 'we can't grow anything even after they have gone'," with the general language of science ("What's that . . . ehm? Is what is called cause and effect . . ."), and this interaction brought the underlying conflict of human rights to the surface. Another interesting element of analysis is the conceptualization of the concept of risk. Starting from an acknowledgment of the lack of precautionary measures ("Girl 2: 'it offers no protection'"), the girls moved onto considering other aspects, such as the "immediacy," "irreversibility" and inequity of the impacts ("we can't grow anything even after they are gone"). In relation to the diagram presented in Figure 8.1., we can observe a "jump" from level 2 to level 3 as the students engaged with the exploration of the context of the issue and the bringing together of facts and value perspectives.

Another example of collaborative work is given here to illustrate students' engagement with language and concepts in science, and particularly the place of scientific knowledge in the controversy:

Girl: Yes well I'm, yes, vegetables are good but not that good
Boy 1: yeah.
Girl: What vegetables have, like, protein in them?
[. . .]
Boy 1: No they have fibre or something like that as well
Girl: Yes but I mean it says here but that um one good, oh Soya beans
    and other protein rich vegetables.
Boy 1: ehm
Girl: what? Ok, so
Boy 1: Um rain is not . . . what is the word?
Girl: Frequent.
Boy 2: Yes . . . not frequent, not consistent enough.

In this excerpt, we can note many of the aspects reported at levels 1 and 2 of the diagram presented in Figure 8.1. Students make interdisciplinary links, drawing on their own personal preconceptions and the reading materials. In addition, the feedback they give to one another during the discussion allows them to grapple with the concept of nutritional value ("good, but not that good") and to begin to build a context for making sense of the arguments in the controversy. Later in the activity, this group's discussion also features a jump of level when they move from the concepts to the discourses of science:

Girl: water pollution along the coast, I am not really sure what that
    means really what water pollution is that?
[. . .]: they are writing things down and looking up info
Boy 2: it's development.
Girl: Sorry?
Boy 2: It is just better development. That is something we would sort of
    believe in.
Girl: Yeah

Negative cases in the analysis also showed that not all groups managed to develop positive ways of working. Data showed a tendency to "solve" conflict, for example, by using language to suffocate disagreement through judgements. In some cases, students adopted a model of leadership that placed emphasis on individual performances and the winning of power over "the other." These relational aspects explain some of the difficulties that students encountered in the third stage of the activity, and more in general, they are valuable cues of the underlying process of knowing that is taking place. These need to be taken into account when evaluating the process of a role-play activity in any particular educational context.

Group Work (Stage 3)—Engaging with Dialog in Conflict

In this excerpt, students progressively approached the conflict of interest and human rights by enacting and questioning contradictory discourses. In particular here, the discourse of economic growth is a particular matter of concern for the students:

> *Jeganatthan (girl 2):* we need, we want the goal to be the benefit of the Indian people not to benefit other countries.
> *Power (boy 2):* No I don't quite agree, [we need lots of prawn farms you see].
> *Power (boy 2):* you see, we actually need lots of prawn farms to boost the economy.
> *Jeganatthan (boy2):* Why? Why do we need more!

In some groups, resolution came after the groups began to address issues of common needs and equity within a scenario of finite resources. This required a series of steps, from exploring personal motivations to decontextualization (i.e., finding common problems in other countries) and problem solving. In relation to the diagram in Figure 8.1., the third part of the activity cast light on the potential for transdisciplinary links based on creativity and sociocognitive and socioemotional involvement in *what is yet to be known* (Varela et al.1993; emphasis added).

CONCLUSIONS

This chapter sought to illustrate the use of role-play as a tool for aligning science education with the changing landscape of science–policy interactions and the emerging epistemology of sustainability science. In such a context, the methodological framework for role-playing allowed for meaning-making, discussion of value assumptions, as well as disciplinary views to develop awareness of the bounded nature of our knowledge and the need for making connections among processes and phenomena in order to grasp the complexity of natural systems. This scenario appears to be in contrast with traditional context of science learning, in which curriculum and assessment are set out to predefine pupils' learning and interactions are narrowed to question–answer responses. During the role-play, new knowledge was being produced while participants confronted themselves with the epistemological and normative categories they held. Hence, "learning" was investigated as the complex and unfolding interplay and interrelationships between learners, action of inquiry, context and environment (Fels 2005).

In line with Kincheloe and Berry (2004), who and how we come to be in relationship with others and our environment is a fluid, interactive process as a continuous back and forth between our perception and the reality in which we are immersed. This has important implications for sustainable education. For example, we can go beyond the false perception that reality can be controlled, or that we can put the other in predefined, univocal categories[6] and separate ourselves from others. Rather, reality takes on a dynamic form and our understanding of the world and ourselves is deepened through the relationships we establish.

From the point of view of language structures, the setting of the role-play allowed for many different language registers and discourses to come together to support the mutual inquiry of the students. As we saw in Chapters 4 and 6, for example, a reflection on language can make us aware of how we see the natural world and the links between our position of inquirers and the knowledge that is being created. Contrary to most traditional settings, in which the voice of the teacher legitimizes knowledge and meaning, in the role-play, the voices of the students came together to formulate meanings and understandings. Through the process of group interaction and building of consensus, we could see embryonic forms of engagement with a different culture of learning and understanding the world. In relation to classroom learning, this kind of reflection can make us more conscious not only of how we teach (What content do I select? What epistemology do I perpetuate?), but also how we learn about the natural world. Knowledge, in that case, is not disconnected from understanding. As reported by Sterling (2002):

> It appears possible to escape the trap of self-reference through a process of meta-learning. A metaview of our thinking is achieved through meta-learning, and beyond this, epistemic learning, which means realizing the nature of our cultural paradigm and moving towards an expanded alternative (4).

In sum, the role-playing activity could be seen from different perspectives. Just as a teaching tool (i.e., to learn about the issue of prawn farming) or as a means for participation, engagement and development of new competences. Such competences derive from a broader understanding of learning, as continuous and shared process, varied in its forms and purposes.

If we are moving toward a situation of increasingly blurred boundaries between science and policy, with a recognition of the limitations of the notion of scientific objectivity for informing rational policy, then it becomes important to look at a different way of teaching, in which the power relations between science and public, teacher and students, school and community are redefined.

NOTES

1. A paper published by Naylor and other nine authors ("Effect of aquaculture on world fish supplies," *Nature* 405, 1017–24, [2000]) provoked a harsh reply from various other researchers, e.g. James H. Tidwell and Geoff L. Allan. "Fish as food: aquaculture's contribution ecological and economic impacts and contributions of fish farming and capture fisheries." *EMBO Reports* 2 (11), 958–63 (2001).
2. The methodological aspects related to the research will not be discussed in detail here, although some links will be provided in the concluding section

to action-research as a reflexive methodology. For similar discussions, the reader can refer to Colucci-Gray (2007) and Fels (2005).

3. With reference to the diagram presented in Figure 8.1., in this stage of the activity, levels I (individual) and II (small group) were subsumed in order for the group to be able to overcome some of the cognitive and emotional conflict that could potentially come from the topic of the role-play by means of discussing and building internal consensus. In this case, the empathetic process was linked to the social dynamics within the group and students' abilities to support one another in understanding their character.

4. In the light of the discussions conducted earlier on prawn farming as an exemplar case, the choice of characters for this particular role-play allowed for a less realistic representation of reality. For example, it did not show contrasting views held by scientists of similar or different disciplines or the controversies that originated among the experts working at the Food and Agriculture Organization. Hence, for such characteristics, the role-play featured a reduced level of complexity, emphasizing the views of the characters and their interests, rather than the epistemological discussions. However, this should not be taken as a shortcoming of the activity, but as a feature of the flexibility of this tool. The result of the activity is not predefined at the beginning, but only the initial, facilitating conditions can be described. The outcome will depend very much on the context of interaction and self-expression, which is being created by the students during the course of the simulation. It will be the task of the teacher to evaluate such process in the light of what the students have gained.

5. Jegannathan and his wife Krishnammal have been recently (October 2008) conferred the Right Livelihood Award, an international award also know as "alternative Nobel Price" " . . . for two long lifetimes of work dedicated to realising in practice the Gandhian vision of social justice and sustainable human development, for which they have been referred to as 'India's soul'."

6. Amartya Sen (2006) pleads for "a recognition that identities are robustly plural" (19), as a sense of belonging and loyalty among human beings come from different groups generated as a result of birth in a particular family, neighborhood association, affiliation with professional groups, religious communities and political alliances. These plural identities can be either "contrasting" or "non-contrasting," depending on the conflicting or complimentary nature of the priorities and demands of different identities (29).

BIBLIOGRAPHY

Adams, D. M. 1973. *Simulation games: an approach to learning.* Worthington, OH: Charles A. Jones Publishing Company.

Barab, S. S., and W-M. Roth. 2006. Curriculum-based ecosystems: supporting knowing from an ecological perspective. *Educational Researcher* 35 (5): 3–13.

Berkes, F., and C. Folke. 1998. Linking social and ecological systems for resilience and sustainability. In *Linking social and ecological systems,* ed. F. Berkes, C. Folke, and J. Colding, 1–27. Cambridge: Cambridge Univ. Press.

Billig, M. 1996. *Arguing and thinking. A rhetorical approach to social psychology.* Cambridge: Cambridge Univ. Press.

Bunting, S. W. 2007. Regenerating aquaculture—Enhancing aquatic resources management, livelihoods and conservation. In *The sage handbook of environment and society,* ed. Pretty et al., 395–410. London: Sage.

Chambers, R. 1997. *Whose reality counts? Putting the last first.* London: Intermediate Technology Publications.

Colucci, L., and E. Camino. 2000. *Gamberetti in tavola: un problema globale.* Gioco di ruolo sugli allevamenti intensivi di gamberetti in India. Torino, Italy: EGA.

Colucci-Gray, L. 2007. An inquiry into role-play as a tool for dealing with complex, socio-environmental issues and conflict. PhD diss., The Open Univ.

Colucci-Gray, L., E. Camino, G. Barbiero, and D.S. Gray. 2006. From scientific literacy to sustainability literacy. An ecological framework for education. *Science Education* 90 (2): 227–52.

Davies, J. 2003. Strategies for building a democratic peace. In *Second track/citizens' diplomacy*, ed. J. Davies and E. Kauffman. London: Rowman and Littlefield Publishers.

Dimitrov, R. 2003. Knowledge, power and interests in environmental regime formation, *International Studies Quarterly* 47: 123–43.

Ellington, H. 2000. Games and simulations—Media for the new millennium. In *The international simulation and gaming research yearbook: Simulation and games for transition and change*, Vol. 8, ed. D. Saunders. London: Kogan Page.

Fels, L. 2005. Complexity, teacher education and the restless jury: pedagogical moments of performance. *Complicity: An International Journal of Complexity and Education* 1 (1): 73–98. Also available online at www.complexityandeducation.ca

Folke, C., F. Berkes, and J. Colding. 1998. Ecological practices and social mechanisms for building resilience and sustainability. In *Linking social and ecological systems*, ed. C. Folke, F. Berkes, and J. Colding, 414–36. Cambridge: Cambridge Univ. Press.

Gallopin, G. 2004. Sustainable development: epistemological challenges to science and technology. Background paper prepared for the Workshop on "Sustainable Development: Epistemological Challenges to Science and Technology," ECLAC, Santiago de Chile, October 13–15, 2004.

Galtung, J. 1996. *Peace by peaceful means.* London: Sage.

Gee, J. P. 1999. *An introduction to discourse analysis: Theory and method.* New York: Routledge.

Kabat-Zin, J. 2005. *Riprendere i sensi.* Milano: TEA.

Kincheloe, J., and S. Berry. 2004. *Rigour and complexity in educational research. Conceptualising the bricolage.* Buckingham: Open Univ. Press.

Mercer, N. 2000. *Words and Minds. How we use language to think together.* London: Routledge.

Mitchell, G. R. 2000. Simulated public argument as a pedagogical play on worlds. *Argumentation and Advocacy* 36:134–50.

Moscovici, S., and W. Doise. 1991. *Dissensions et consensus. Une theorie generale des decisions collectives.* Trans. Pina Lalli as *Dissensi e Consensi.* Bologna, Italy: Il Mulino.

Osborne, J. 2003. The role of argument in science education. In *Research and the quality of science education*, ed. K. Boersma, M. Goedhart, O. De Jong, and H. Eijkelhof, 367–80. Dordrecht, The Netherlands: Springer.

Potter, J. 1996. Discourse analysis and constructionist approaches: theoretical background. In *Handbook of qualitative research methods for psychology and the social sciences*, ed. J.T.E. Richardson. Leicester: BPS Books.

Primavesi, A. 2000. *Sacred Gaia: Holistic theology and Earth system science.* New York: Routledge.

Sen, A. 2006. *Identity and violence—The illusion of destiny.* New Delhi: Penguin Books.

Sterling, S. 2002. A baker's dozen—Towards changing our "loaf." *The Trumpeter* 18(1):1–14.

Van Ments, M. 1999. *The effective use of role-play. Practical techniques for improving learning*, 2nd edition. London: Kogan Page.

Varela, F., E. Thomson, and E. Rosch. 1993. *The embodied mind: Cognitive science and human development*. Cambridge, MA: MIT Press.

Conclusions
Complexity, Emancipation and Empowerment

Donald Gray, Laura Colucci-Gray, Elena Camino

This book started with a question about the role played by science and education—two major sectors of the life of our Western societies—in the current critical scenario of global environmental changes. The analysis of current issues and debates surrounding complex and controversial environmental issues has provided the backdrop for a reflection on the modes of production and use of scientific knowledge. The recognition of the complexity of coupled systems—human and natural—and the environmental emergencies challenged the image of technoscientific enterprise. Science as a body of consolidated knowledge, to which to refer for the solution of problems and for the fulfilment of people's requirements, has fallen short. Some authors have called for holistic approaches to be developed, including interdisciplinary methods:

> a core sustainability science research program has begun to take shape that transcends the concerns of its foundational disciplines and focuses instead on understanding the complex dynamics that arise from interactions between human and environmental systems (Clark 2007, 1737).

Besides, technoscience has emerged as a cultural process that is deeply embedded and interwoven with the economic and political forces that drive modern societies. Such a new role of science raises questions about the relations between scientific endeavor and public choices, and about the actions to undertake in order to preserve the quality of the democracy (Fazzi 2008).

In this framework, education is put into question. In the face of the recognition of the changes that are expected from both science and society, confronted by the unsustainability of human actions, science education is called upon to complete a double task: one is to revise its own aims, methodologies, content and assessment criteria to make them more commensurate with a new epistemology, which recognizes the uncertain, multiple and transitory nature of human knowledge; the other one is to cope with the new and pressing requirements of a participatory democracy.

This book has endeavored to recount the journey, the reflections and the actions of a group of people who have been involved in developing, trying out and promoting educational contexts aimed at integrating the multiple dimensions of learning—cognitive, social, emotional, spiritual—in a process of personal and collective searching. Specifically, scientific knowledge—in our view—is no longer to be considered a body of notions, but rather a process of continuous interaction and integration among multiple, dynamic and partial views, which are subjected to reflection, exchange and reframing, both at an individual and a collective level. Engaging with science can become a way of gaining insights into possibilities and approaches for seeing the world, which is at the center of a conscious and dynamic movement between the processes and the products of an inquiry. In this context, science education can contribute toward a sustainable education by putting emphasis on the awareness of the complexity of natural systems; the characteristics of ignorance, contingency, impermanence and uncertainty of human knowledge and portraying and promoting attitudes of humility and choices of reversibility in human actions. This more fluid perspective on knowledge can accompany emancipation and empowerment of citizens toward power structures, allowing them to formulate new cultural constructs beyond what is being currently perpetuated by the institutional hierarchies in the political, economic and educational realms. In this regard, assessing the quality of the process of knowledge construction can be an important tool for progressing along the path of conscious learning. As a means of example—and we reserve to another context the task of dealing with the important aspect of assessment—such criteria of quality may be concerned with the process of knowledge construction (the possibility to participate; to include, express and listen to multiple perspectives; the disposition toward the discussion of contradictions and paradoxes), as well as with the knowledge product itself (the relevance, affordability and intelligibility of the knowledge products for the people involved). In this view, science education can become a vehicle of substantial changes in the practice and representation of civic spirit. Emancipation and empowerment can be associated with an environment of commitment, conviviality and creativity—making society, schools and living environments more hospitable and sustainable.

BIBLIOGRAPHY

Clark, W. C. 2007. Sustainability science: A room of its own. PNAS 104 (6): 1737–8.
Fazzi, L. 2008. La partecipazione fra teoria e pratica: nodi metodologici, sfide e scenari future. In *Educare alla cittadinanza partecipata,* ed. L. Mortari, 141–69. Milano: Mondadori.

Contributors

Marta Angelotti is a contract lecturer in the Initial Teacher Education and undergraduate natural sciences degree programs at the University of Turin. She is a member of the Research Group in Science Education in the Department of Animal and Human Biology and academic secretary of the Interdisciplinary Research Institute on Sustainability in Turin University.

Giuseppe Barbiero is a lecturer of ecology and biogeosciences in the Faculty of Education at Aosta Valley University. He is chief of *The Project IDEE (Intelligences for Deep Ecology Education)* at Aosta Valley University. His main research interest is the biophilia hypothesis, particularly the role of wilderness fascination in the restoration of mental fatigue. Other research interests are the Gaia hypothesis, focused on the living state of matter and symbiotic systems; and environmental risk assessment of transgene introgression by genetically modified crops.

Alice Benessia is a graduate student in physics, a professional photographer and a doctoral student at the University of Padova. She has lived and worked in New York for several years while completing a Master of Fine Arts at the School of Visual Arts of New York (2002), and a Master in Philosophical Foundations of Physics at Columbia University in New York (2003). She is an executive member of the Interdisciplinary Institute for Research on Sustainability in Turin University.

Fabrizio Bertolino is a lecturer in general and social pedagogy at the University of Val d'Aoste, and is a member of the Interdisciplinary Research Institute on Sustainability in Turin University. He holds courses and laboratories for the degree courses in educational studies and primary education. His main areas of research include environmental and sustainability education in teacher education and the role of natural environment in children's development.

Elena Camino has a degree in physics and is a senior lecturer in the Department of Animal and Human Biology at the University of Turin. She is a co-founder of the Research Group in Science Education (University of Turin) and a co-founder of the Interdepartmental Centre IRIS (Interdisciplinary Research Institute on Sustainability).

Laura Colucci-Gray has a degree in natural sciences and a PhD in science education. She is currently a lecturer in the School of Education at Aberdeen University, and is also member of IRIS. Her research interests focus on teacher education, as part of the Scottish Teachers for a New Era initiative, and sustainability education. Her interest is primarily in the use of interactive pedagogies, such as role-plays, to deal with socio-environmental issues and conflict.

Martin Dodman is a graduate in literature and language at The College of Arts and Technology at Cambridge University. He carried out postgraduate research and study in comparative linguistics and literature at Columbia University in New York. Currently, he works as a lecturer in comparative education and didactics of language in the Faculty of Educational Science of the Free University of Bolzano. He is also a freelance consultant in education and educational research.

Luca Giunti holds a degree in natural sciences from the University of Turin, and is a park ranger in the Natural Park of Orsiera-Rocciavre' in Piedmont. He has a long-standing experience of collaboration with the Science Education Research Group at Turin University, and has lectured in undergraduate and postgraduate courses on sustainability.

Donald Gray is a senior lecturer in the School of Education at the University of Aberdeen. He holds a degree in zoology and a PhD in science education. He has a background in environmental and outdoor education, as well as science education. He is currently the research director for the Scottish Teachers for a New Era initiative at the University of Aberdeen. He is a member of IRIS.

Ângela Guimarães Pereira has a PhD in environmental assessment from the New University of Lisbon. She has worked at the Joint Research Centre of the European Commission since 1996, where, with others, she formed the Knowledge Assessment Methodologies group. She is responsible for activities on science and society interfaces that range from knowledge quality assurance methodologies to social research and deployment of new information technologies for European Union inclusive governance prospects.

Daniela Marchetti has a PhD from the Department of Animal and Human Biology at the University of Turin, and is a member of the Research Group in Science Education. She has delivered courses in biological and sustainability education at the University of Aosta. Her research focuses on interactive activities and learning processes in sustainability education.

Anna Perazzone is a lecturer in science and environmental education in the Department of Animal and Human Biology. She is a member of the Research Group in Science Education and the Interdisciplinary Institute for Research on Sustainability at Turin University. She specializes in primary education and initial teacher education, with a focus on didactics of biology and environmental education.

Stephen Sterling is the Schumacher Reader in Education for Sustainability at the Centre for Sustainable Futures at the University of Plymouth; Senior Advisor for education for sustainable development to the UK Higher Education Academy; and a visiting research fellow at the Centre for Research in Education and the Environment at the University of Bath. His research interest is in the interrelationship between learning, ecological thinking and systemic change toward sustainability.

Marco Davide Tonon is a lecturer in the Department of Earth Sciences at Turin University, and is also the director of IRIS. His main area of teaching and research is didactics of the earth sciences, and he teaches on courses and laboratories for the degree course in primary education at Turin University.

Index